Economic Evaluation of Transport Projects

Ginés de Rus · M. Pilar Socorro · Jorge Valido ·
Javier Campos

Economic Evaluation of Transport Projects

 Springer

Ginés de Rus (ID)
Department of Applied Economic Analysis
University of Las Palmas de Gran Canaria
Las Palmas de Gran Canaria
Spain

Departament of Economics
University Carlos III of Madrid
Getafe, Spain

FEDEA
Madrid, Spain

Jorge Valido (ID)
Department of Applied Economic Analysis
University of Las Palmas de Gran Canaria
Las Palmas de Gran Canaria
Spain

M. Pilar Socorro (ID)
Department of Applied Economic Analysis
University of Las Palmas de Gran Canaria
Las Palmas de Gran Canaria
Spain

Javier Campos (ID)
Department of Applied Economic Analysis
University of Las Palmas de Gran Canaria
Las Palmas de Gran Canaria
Spain

ISBN 978-3-031-35958-3   ISBN 978-3-031-35959-0 (eBook)
https://doi.org/10.1007/978-3-031-35959-0

This Springer imprint is published by the registered company Springer Nature Switzerland AG
The registered company address is: Gewerbestrasse 11, 6330 Cham, Switzerland

Preface

The purpose of this book is to show how the social appraisal of public policies can be very helpful for the improvement of public intervention in the economy. The economic judgement of whether an investment project or a new regulation is socially worthy is based on the comparison of the benefits and costs of public intervention. We are convinced that the set of cost–benefit analysis general equilibrium rules is the best approach for the social appraisal of most of the projects in the transport industry. A key idea is to distinguish between impact and net social welfare; another is to understand that the full potential of cost–benefit analysis can only be obtained with the right institutional design. Many of the criticisms concerning the limited influence of cost–benefit analysis in actual policy decisions do not distinguish between the method and the institutional framework in which the evaluation takes place. In this book, we develop a simple model where the reader can see the derivation of the practical rules for the assessment of transport projects. Then, the cost–benefit analysis rules are used for the evaluation of an investment project in rail infrastructure, for the decision on the transport network to be constructed, and for the analysis of a public policy consisting of subsidizing air transport. Both cases show the power of economics to realize when it is socially desirable to spend public money on a particular intervention and also to choose between alternative ways to achieve the same target.

The research presented in this book draws on the technical report commissioned to the authors by the Independent Authority of Fiscal Responsibility (AIReF) for the Spending Review on Transport Infrastructure. It was funded by the EU via the Structural Reform Support Programme (SRSS). It reflects only the authors' views and does not imply a policy position of the European Commission or the AIReF, neither are responsible for any use that may be made of the information it contains. We are deeply indebted to Per-Olov Johansson and Bengt Kriström for their advice and encouragement. Finally, we are also grateful to the staff at Springer, specially to Sharmila Anbu, for their advice and support during the publication process. Any remaining errors are our sole responsibility.

Las Palmas de Gran Canaria, Spain   Ginés de Rus
                                                                     M. Pilar Socorro
                                                                        Jorge Valido
                                                                      Javier Campos

Contents

Chapter 1
Introduction

A project, interpreted as any public policy affecting the economy, should not be approved unless the intervention is, not only expected to generate social benefits above the social costs, but with a net result superior to the next best alternative. In summary, the social welfare change should be positive considering all the relevant effects on the economy. This implies a rigorous exercise of identification, measurement, and valuation of the flow of benefits and costs of the ways to deal with the problem at hand so that the analyst can provide the government, and society, with an estimation of the consequences of public intervention on social welfare. This is what cost–benefit analysis (CBA) is about.

An investment in infrastructure or a public policy changing the regulation in a mode of transport has consequences beyond the primary market, directly affected by the intervention, but on many other markets through different ways. First, on the direct users of the transport services; and then, in secondary markets with products that are complements and substitutes of markets affected by the change in transport costs due to the project, including changes in modal split, externalities, and other second-order effects. Even the income multiplier effect on the economy could be significant under conditions of involuntary unemployment.

A key point to avoid misunderstandings concerning the purpose of CBA is the distinction between impact effect and net welfare effect. Before the approval of the project, the social appraisal of the intervention requires a clear distinction between the impact on gross value added, employment, and so on, and the net welfare effect of the project. The income multiplier effect of the project is a good example of an actual impact of public expenditure, but is irrelevant in the calculation of the net social benefit when the multiplier is similar to the next best alternative.

The evaluation must be carried out before the approval of the policy. The socially useful analysis is essentially ex-ante, comparing, based on the available information, the benefits and costs that are expected with the public policy under evaluation. To guarantee an unbiased appraisal, the practitioner in charge of this task should be indifferent in his preferences regarding the project. This requires a neat independent evaluation process with respect to the government and interest groups. The reason is

© The Author(s), under exclusive license to Springer Nature Switzerland AG 2023
G. de Rus et al., *Economic Evaluation of Transport Projects*,
https://doi.org/10.1007/978-3-031-35959-0_1

very simple: the decisions corresponding to the maximization of social welfare may not have to be aligned with those corresponding to the maximization of the probability of reelection, or the goals of some interest groups affected by the intervention. For this reason, institutional design is at least as important as the methodology (see Engel et al., 2014).

Public officials and analysts have different tools to deal with this task. Our first option is CBA, the way of thinking of the economist, as the best tool available for the social appraisal of transport improvements, a field where CBA has been widely applied, producing a sound body of theory and evidence. It is the method that we have at our disposal to assess the net welfare changes of projects and public policies resulting from the intervention compared with the counterfactual. Firstly, it requires the establishment of an analytical framework in which the problem is identified, all feasible alternatives to solve it are considered, all groups significantly affected by the intervention count, and the appraisal is done respecting the individuals' preferences and expressing the effects in monetary units.

There are other tools for the evaluation of transport projects, like multi-criteria analysis and computable general equilibrium models (CGE). The first one does not measure changes in social welfare, and the second one is more appropriate for large shocks, such as macroeconomic shocks in the economy. When CGE is used for the social appraisal of projects, like the construction of a new road, the use of an existing CGE model designed for large economic impacts is not going to add any value to the CBA of the project unless a specific CGE model is developed (see de Rus, 2023; Kriström, 2023). A standard CGE model built to capture the effects of trade liberalization will barely capture differences between the net welfare effects of urban commuting or high-speed rail (HSR) projects. Both projects will have induced effects in the economy following the initial investment, but their direct effects and possible wider economic benefits (WEBs) are substantially different (see Laird & Venables, 2017).

The justification of projects based on the impact on the economy, adding multiplier effects and the like, is bad practice in CBA. In many cases, investment in transport infrastructure is often justified by politicians as a sufficient condition for economic development. Laird et al. (2014) warn of the use of expenditure and costs instead of genuine benefits. They mention the recent shift by planners in the UK, using changes in gross value added, including wages, as a benefit.

Transport projects are usually evaluated using the derived demand. This demand summarizes valuable information on the effects of the reducing transport costs on the economy. The use of the well-known 'rule of a half' for the measurement of the benefits of transport projects has been confused with a narrow partial equilibrium approach where the practitioner is unduly assuming everything constant in the rest of the economy. Moreover, the use of market demand functions for general equilibrium welfare effects assessment is justified, under some conditions. Using reduced-form elasticities, it is possible to assess the welfare effects of projects as they incorporate general equilibrium effects in all the affected markets (Chetty, 2009; Just et al., 2004; Kleven, 2018).

Politicians, unsatisfied with the net welfare effect approach, use to argue that public transport infrastructure investment delivers other benefits beyond direct user benefits, such as changes in productivity and industry reorganization. The problem is that, in principle, those gains in productivity are not additional and have already been measured with the transport demand (Mohring & Williamson, 1969). Some projects have WEBs, but some distortions are required in the rest of the economy, like the endogeneity of the wage gap in the case of agglomeration economies following changes in proximity and labour density (Venables, 2007), or the benefits of urban redevelopment in the presence of a market failure (Laird & Venables, 2017).

It is possible to measure the social benefits of a transport project with the market demand in the primary market in the absence of distortions in the rest of the economy (Harberger, 1965; Johansson, 1993). Even with distortions in secondary markets, it is unnecessary to include indirect and induced effects when these effects are expected to be of the same order of magnitude in the next best alternative. For relatively small projects, the observed general equilibrium demand is usually sufficient to assess the expected welfare effect of public intervention.

This book is about CBA of policies and investment projects in transport (see Mackie et al., 2014), but is neither a textbook nor a comprehensive survey. The purpose of the book is to stress the importance of having an analytical framework, a model where the practical criteria of measurement are obtained. CBA is the appraisal methodology most frequently used in the main supranational and national guidelines for the economic evaluation of investment projects. A quick revision of the guidelines of the European Investment Bank, the European Commission, or the Asian Development ment Bank; as well as some countries with an evaluation tradition, such as the UK, the Netherlands, or Sweden, shows they share a common body of rules of thumb that can be applied to standard projects. The warning is to avoid the mechanical transfer of these rules, sometimes mixing different sources, with the risk of leading to inconsistencies, double counting, and other errors that bias the results.

Therefore, the starting point is to have a model that establishes a rigorous analytical evaluation framework, with explicit assumptions, and from which practical evaluation rules are formally derived (see Johansson, 1993; Johansson & Kriström, 2016). This is the objective of this book, to present a basic model for the evaluation of transport projects and policies and to apply the rules to an investment project and a public policy.

CBA is based on the monetary valuation of changes in individual well-being caused by the implementation of a project compared with the counterfactual, to assess whether society is better-off with such a public intervention. The incremental change in individuals' welfare responds to the valuations made by the affected parties, directly (stated preferences) or through the observed behaviour in the market (revealed preferences). Economists use willingness to pay and willingness to accept to measure the benefits and costs of public policies. Although it is not possible to measure the change in individual well-being because it is not observable, it is possible to obtain the monetary valuation of that change.[1]

Measurement of the change in individual utility (well-being) according to private valuations is the previous step to assess the change in social welfare. We proceed as if the government maximizes social welfare,[2] which can obey different functions but usually responds to the following four properties (Mas-Colell et al., 1995, p. 825): (i) Non-paternalism. In the expression of social preferences, only individual utilities matter. (ii) Paretian property. Welfare increases with the utility of each individual. If one individual is better-off without making anyone else worse-off, there is an increase in social welfare. (iii) Symmetry. In the evaluation of social welfare, all individuals are on the same footing. (iv) Concavity. This is based on inequality aversion. The extent of compensation is determined by the degree of inequality in society.

The social welfare function, therefore, depends on the well-being of the individuals, who try to maximize their utility according to their preferences, prices and income. The utility of individuals is a function of the goods and services they consume, whose prices and quantities change with a public intervention that modifies the equilibrium in the economy, affecting them as consumers, owners of the factors of production, taxpayers, and those affected by changes in externalities. CBA attempts to assess the effect of government intervention on social welfare, including all social groups.[3]

This evaluation procedure is not only useful ex-ante, but also when the project has been completed (ex-post) or when it has been running for a sufficiently long period of time. In these last two cases, though the evaluation does not influence whether the project should have been approved or rejected, it may help to introduce modifications and, definitively, allows to improve future evaluations.

The usual approach in the ex-post evaluation is to redo the calculations that were done in the ex-ante CBA but with observed values. Although this procedure provides a good deal of information about the actual economic effects of the projects and how these effects compare with the ex-ante analysis, its weakness is that the evaluation is carried out according to the same model as the ex-ante evaluation.

Another method, based on causal inference, can be followed to carry out the ex-post evaluation. It consists of applying statistical models to the data observed before and after the transport interventions, trying to estimate the incremental economic effect of the project. It is an empirical method and, although it gets around the problems of CBA mentioned above, it has to overcome other statistical problems to adequately reveal causal effects. Both methods pursue the same goal: to understand the impact of public intervention on the economy through investments and transport policies. The results of the ex-post econometric analysis feed and reinforce the ex-ante CBA.

---

[1] For cases where the individual' preferences are distorted and, hence, they are not a good guide for individual' well-being (see Adler & Posner, 2001).

[2] Even if this is not the case, CBA is still socially useful under the appropriate institutional design.

[3] For an analysis of who stand in CBA (see Johansson & de Rus, 2019, and Zerbe, 2018).

When public officials calculate the net social benefit of a project and obtain a positive result, most of the time they are applying the criterion of potential compensation, which implies that, if the redistributive consequences of the project were sufficiently regressive, it might happen that the conventional net present value does not reflect the real impact on welfare or, what is worse, that its sign changes. What is the way out of this problem?

In practice, the potential compensation criterion is usually applied under the assumption that the fiscal policy has mechanisms to successfully redistribute income. Its application is also justified by arguing that since the government carries out many projects, in the long-term, the society will be better-off. It should not be forgotten that, as in the case of expropriation, the criterion of potential compensation is accompanied by somewhat imperfect compensation that mitigates the losers' damage. It could also be argued that the difficulties of identifying the final beneficiaries and losers in many cases can make full compensation impossible; or that, being possible, the costs of identifying winners and losers and establishing compensation mechanisms outweigh the benefits.

This book is on the social appraisal of transport projects. Its purpose is to show how the economic evaluation of transport investments and policies provides valuable information for decision-making beyond intuition or goodwill. Although the methodology is common to many other areas, the characteristics of transport as an economic activity make this field particularly suitable for the application of CBA. Transport improvements resulting from investment projects and policies can be typically considered as exogenous interventions in transport markets, which move the economy from one equilibrium to another, commonly through the reduction of the generalized price (monetary price, time, and other disutility components).

The different modes of transport have experienced some organizational and technological changes in the recent past. A high-speed train over medium distances is closer to air transport than to traditional railway service. The transport and logistics of container handling with the new container ships and the new specialized port terminals use very advanced technology and bear little resemblance to the manual loading and unloading of cargo that stevedores formerly carried out in ports. Despite all the technological changes and the structural differences between the different transport modes, they share some common characteristics from the perspective of economic analysis. Some of them are particularly relevant for the economic appraisal of public intervention in transport markets (see de Rus et al., 2003).

Production Technology: Infrastructure and Services

Two different types of activities can be distinguished within the transport industry. Public and private firms are engaged in the construction and operation of infrastructures (ports, airports, roads, etc.), while other firms and individuals operate the vehicles that use these infrastructures to produce the services supplied to other firms or for their use (shipping companies, airlines, bus operators, private cars).

In all modes of transport, these two elements are present when examining their production function. Differences between transport modes are largely due to technological reasons. Both the characteristics of the vehicles and the infrastructure they require, somewhat affect the type of market organization and the degree of feasible competition. In the case of railways, the management of the infrastructure and the operation of services require a high degree of coordination, which partially explains why railways were vertically integrated in the past. In contrast, roads hardly need coordination between the vehicles using the infrastructure, in the sense that it is not necessary to determine schedules for access to the infrastructure, and hence, the market works with a separation between the companies that are dedicated to the management of the infrastructure and the operators of the vehicles.

The infrastructure side of the transport industry is, in many cases, subject to public provision, and the social appraisal of projects is required for an efficient allocation of public funds.

A Basic Input: Time

The economic analysis of transport activities requires the explicit consideration of an essential input in the production function of transport services: the time of the users (both in the case of passengers and freight). Many decisions of firms and individuals cannot be understood without explicitly including time in the analysis. Time is an input provided by the transport user.

The demand for transport is a derived demand. In general, the user moves between the origin–destination points for other purposes (work, study, leisure). Therefore, the user tries to minimize the generalized cost of the trip, which includes the price and the total value of the time invested by the user. In freight transport, the time invested is also important since the speed and reliability of deliveries are inversely related to storage costs. Many models in transport economics are based on a trade-off between production costs and user costs. The evaluation of transport systems, the configuration of networks, the modal distribution, and the structure of markets requires the consideration of the trade-off between both types of cost.

Transport Supply: Non-storability, Indivisibilities and Sunk Costs

Transport is not storable. Once the operator supplies a certain volume of capacity units, it must be consumed at the time when the service is being produced, or it will be irreversibly lost. This characteristic has important implications for transport managers, which must give an adequate dimension to their supply according to the characteristics of the demand.

Transport infrastructure has limited capacity, measured by the maximum number of vehicles capable to operate at the same time. Infrastructure has discontinuities in the possibility of its expansion (e.g., to increase the capacity of a road, an entire additional lane must be built, or for an airport to accommodate more arrivals and departures of aircraft, a new runway or terminal must be built).

If an infrastructure manager provides unnecessary capacity or a firm puts more vehicles in circulation than necessary, most of the supply will be lost, with consequent waste of resources. But if, on the contrary, the capacity is insufficient, queues, congestion, and lower service quality will be the result.

In addition to the problems of mismatches between supply and demand, the non-storability nature of infrastructure and transport services also has implications concerning the demand profile and the size of the infrastructure and the fleet of vehicles. Transport demand is usually not uniform during the day, neither during the days of the week nor during the period of the year. This means that to avoid problems of scarcity or congestion, operators must dimension capacity for peak periods. Hence, in the off-peak, part of the infrastructure and vehicles will be unused without any savings given the high proportion of fixed or semi-fixed costs in the transport industry.

Infrastructures are specific assets, and their costs are sunk, since the assets cannot generally be used for any other purpose than that for which they were built. On the other hand, these are assets with very long useful life and with a high share of the costs generated in the initial construction phase. Given the long-life span of the assets, the high and sunk investment costs, and demand uncertainty, the economic appraisal of infrastructure projects is essential.

Network Effects

One of the economic characteristics of the transport industry, shared by other industries such as telecommunications or electricity, is the existence of network effects. In the case of transport, these network economies affect both the infrastructure side of the market (physical networks) and the operation of vehicles. In infrastructure, adding a new connection to a network, such as an additional airport or a new road, increases the value of existing assets, increasing the possible combinations. These network effects must be considered when evaluating the benefits of the entry into service of a new infrastructure, and for the design of its capacity, since the number of users who finally use the infrastructure and the social benefits may be higher than in the absence of network effects.

In transport services, some network effects arising from the advantages of a large number of users can also be identified. In particular, in the modes of regular transport—buses, air transport, maritime services—as the number of passengers increases, companies respond by introducing greater frequencies, which allows all users to reduce their waiting times and to have a more convenient schedule.

A dense network allows changes in the productive organization like the 'hub-and-spoke' type, in which there are a series of main nodes (*hubs*) that are connected by large capacity vehicles, and a series of secondary nodes that only have a direct connection with one of the main nodes and that are served by smaller vehicles and lower frequencies. With this type of route structure, companies obtain benefits from the economies of vehicle size, which can be used in segments with high traffic.

Negative Externalities

The construction of the infrastructure and the provision of transport services are associated with negative externalities such as pollution or noise. The unavoidable use of a physical space to locate these assets requires the consumption of natural resources (land, diversion of natural water courses, impoverishment of the landscape, barrier effects for natural habitats, etc.), which must be valued in project evaluation. The use of vehicles produces many other externalities such as pollution and noise. Accidents also fall into this category because, although a large part of the costs generated by an accident is suffered by the agent who causes the accident (personal injury and vehicle damage) or has to pay through the insurance (damages to third parties), there are other costs that he imposes on the rest of society like pain and suffering not covered by legal compensation, police costs, hospital costs, etc.

Congestion is also considered an externality. It is an externality that users of transport services cause each other (and in principle none outside the industry is affected). The externality occurs because each user, when making his decision to use a road, only takes into account the cost to him of the time he will spend on the trip, plus the monetary cost of using the vehicle, but does not take into account that he is making circulation less fluid for everyone. Then, he is imposing a cost in terms of extra time on other road users that the user does not have to pay and, therefore, does not internalize in his travelling decisions. The reduction of congestion is a benefit associated with many projects consisting of the expansion of capacity.

Agglomeration Economies

The location of economic activities matters, and new insights appear once the space is included in the economic analysis of transport improvements. Social appraisal of projects that change proximity and modify initial location decisions by firms and individuals has benefitted from this change of perspective. The analysis of the problems of industrial location, regional development, or formation of cities emerges strongly with the so-called new economic geography in which the concepts of increasing returns to scale and imperfect competition are fundamental to explain the pattern of the spatial location of economic activity. Both concepts play a decisive role, first in the new industrial economics, then in international trade and economic growth theory, and finally in spatial economics (Krugman, 1998).

In the traditional model of perfect competition, the effects of an investment in infrastructure that reduces the cost of transport are well-known: reduction of marginal cost, profits in the short-term, entry of new firms, and new equilibrium in the long-term with a lower price and higher quantities. Consumers are the ultimate beneficiaries of reduced transport costs. Nevertheless, relaxing the assumption of perfect competition, we find positive externalities due to the increase in firms and labour density. The reduction in transport costs increases proximity and may lead to positive externalities when firms locate close to other firms.

There are, for example, economies of agglomeration resulting from a reduction in commuting costs that induce a higher concentration of jobs in the centre. Increased labour productivity associated with higher labour density increases productivity. However, a sound appraisal requires the inclusion of any negative effects in the areas where these workers come from. It is considered that for urban areas, it is more likely that the positive effect dominates. In the case of intercity transport projects, the cost of the periphery could exceed the benefits of agglomeration. The outcome is context-specific, depending for example on land prices and the wage gap between areas (see Duranton & Puga, 2004; Graham, 2007; Venables, 2007).

The Costs of Transport and the Consequences of Pricing Policies

The recovery of transport infrastructure costs does not follow a single pattern, with differences between countries and by mode of transport. The discussion of whether the short-term or the long-term is the appropriate time dimension for marginal cost pricing necessarily leads to the question of whether users are willing to pay for the capacity if the first alternative is chosen. The construction of new infrastructure cannot be evaluated without solving the pricing question, as the benefits and costs depend on the volume of demand and its composition.

In the production of any transport service, there are three types of costs: the producer's, the user's, and the external cost. The decision on who pays and how to pay these costs has far-reaching consequences on efficiency and equity. When external costs are not internalized, the price of using polluting modes of transport or generating high noise levels will be relatively more attractive. It may be decided that it is the taxpayer who finances the construction of public infrastructure, in which case the demand for the use of transport services using the infrastructure will increase.

The discussion about what prices should be charged for the use of transport infrastructures and services, which will determine how the three components of total costs are allocated, remains one of the most controversial, and we can certainly say that it is still far from being resolved. Looking at the three cost components and starting with the external costs, we can see that it is not very difficult to internalize these costs. The use of Pigouvian taxes is a way for prices to regain their role as signals for the efficient allocation of resources. This also applies to the non-internalized part of the user's costs in the case of congestion.

Transport producer's costs include costs of building, maintaining, and operating the infrastructure, and other fixed and variable costs incurred by having a fleet of vehicles in operation. For transport service companies operating in competitive markets, prices will tend to marginal costs in the long-run. In the case of infrastructure, there are differences. What price should be charged for the use of the road by a heavy vehicle transporting goods? What price should a traveller pay for a HSR service? Economists agree on marginal cost pricing, and if costs are not covered and there are budgetary constraints, prices should deviate from marginal costs with the least possible loss of efficiency.

Should we recommend the use of short-run marginal cost or long-run marginal cost? Here, the unanimity of economists disappears, and the complexity of transport pricing problems with indivisibilities, sunk costs, long asset life, and economies of scale appears in all its harshness.

In a world where actual infrastructure capacity will rarely be optimal, the decision on whether to opt for short-run or long-run marginal cost pricing has far-reaching economic consequences in terms of the modal distribution of traffic and who will pay for transport costs. It is also argued that for certain modes of transport, such as the metro or urban buses, fares should be lower than marginal cost and, not only as a second-best policy to alleviate congestion and pollution problems but also as an equity policy. Governments often deviate from marginal cost pricing, with arguments based on equity criteria. Transport is a necessary service for all individuals, and the supply of public transport services is essential, especially for low-income households. Although equity is based on other principles, the economic analysis of projects dealing with an equity problem can also help to choose the most efficient alternative to achieve the same equity goal.

The content of this book is the following. Chapter 2 presents the analytical framework for the social appraisal of any public intervention in transport. The analysis is based on a simple model from which practical rules for the measurement of benefits and costs are derived. Chapter 3 applies this methodology to the evaluation of railway investments, evaluating a HSR project, a mode of transport which is characterized by a high proportion of sunk costs and where the volume of demand, the time saving and its value, and the modal split changes in the corridor are crucial for the economic profitability of the project. Chapter 4 analyses the relationship between pricing and investment and its effects on economic evaluation and public decisions when there are different transport alternatives. Pricing affects demand and, hence, social welfare. The social profitability of the project can be quite different depending on the pricing policy. A particular charging scheme may favour the creation of a specific transport infrastructure network, leading to irreversible long-term equilibria that would not be optimal under other charging schemes. Finally, in Chap. 5, CBA is applied to the evaluation of a transport policy. We analyse the efficiency of *ad valorem* and specific subsidies for resident passengers in air transport markets. This chapter shows the potential of the economic analysis to understand the consequences of alternative options to achieve the same target and how this evaluation provides a solid foundation for the selection of the best mechanism.

Chapter 2
Cost–Benefit Analysis of Transport Projects: Theoretical Framework and Practical Rules

2.1 Introduction

Transport improvements can be typically contemplated as exogenous interventions in transport markets, which move the economy from one equilibrium to another, commonly through the reduction of the generalized price (composed of monetary price, time, and other disutility components) borne by transport users. Even projects aimed to provide new capacity and, of course, transport policies such as subsidies to passengers, can be analysed as a reduction in the generalized price of transport. Although there are different reasons that, in principle, could justify these public interventions (e.g., increasing accessibility, improving safety, decreasing congestion and scarcity, or reducing negative environmental externalities), the question is not whether there are social benefits from public intervention, but whether these potential benefits are large enough to offset the opportunity cost of the resources diverted from other uses to obtain those benefits. This is the challenge of the economic evaluation of projects and policies, whose main objective is to assess changes in the well-being of individuals directly or indirectly affected by their implementation.[1]

Social welfare changes can be approximated through the sum of the changes in the surpluses of the agents affected by the project, or through the sum of the changes in willingness to pay (WTP) and in the use of real resources, ignoring income transfers. The measurement of the effect of a reduction in the generalized price of transport is different in the case of a unimodal or multimodal analysis. This chapter follows Mohring (1971, 1993) and covers alternative ways of dealing with the economic evaluation of transport projects, explaining the rationale of the different options. Although the alternative methods aimed to measure the change in social welfare are equivalent, when practical rules are not supported by a robust theoretical framework, there is a risk of underestimation or overestimation of the net social benefits of projects due to omissions or double counting.

---

[1] This chapter is based on de Rus et al. (2022).

© The Author(s), under exclusive license to Springer Nature Switzerland AG 2023
G. de Rus et al., *Economic Evaluation of Transport Projects*,
https://doi.org/10.1007/978-3-031-35959-0_2

The cost–benefit analysis (CBA) of transport projects is commonly conducted using a few measurement rules that are simply obtained from changes in prices and quantities of the derived demand for transport. We follow the theoretical framework developed by Johansson (1993) and Johansson & Kriström (2016), and for the measurement of the effects of transport projects, we follow Johansson & de Rus (2018) and de Rus & Johansson (2019). This chapter develops a model that includes the explicit consideration of time in the generalized prices of goods and services, and the corresponding budget constraint given the time endowment of the individual, with the emphasis on the rigorous derivation of the CBA rules.

There are other economic impacts linked to the response of the private sector to changes in proximity, and these effects could be significant in specific contexts affecting productivity effects and land development. Nevertheless, the main components of transport projects are those accruing to users, and only when prices deviate from marginal costs in secondary markets, other benefits not fully internalized like agglomeration economies should be considered (Laird & Venables, 2017; Mackie et al., 2011). Although the chapter also briefly discusses these project effects, the main target is the analysis of the alternative ways to deal with the CBA of public interventions and the derivation of consistent rules for the practical assessment of projects.

Although we follow the conventional CBA methodology, the analysis in this chapter is based on the model presented in Sect. 2.2, which aims to derive the fundamentals of these measurement rules, explain their assumptions, and clarify the conditions under which they hold. This helps to derive consistent criteria for project evaluation, avoiding pitfalls such as double counting. Section 2.3 discusses some CBA rules and policy recommendations, as well as the equivalence between the aforementioned alternative approaches to project appraisal. In this context, the use and misuse of shadow prices are also discussed, as well as a brief discussion of indirect effects and the so-called wider economic benefits (WEBs). Prices affect demand and, thus, the social surplus of transport investments. For this reason, in Sect. 2.4, we discuss the relationship between pricing and investment. Finally, Sect. 2.5 summarizes the main conclusions of this chapter.

2.2 Cost–Benefit Analysis of Transport Projects: Theoretical Framework

2.2.1 A Basic Model to Measure Social Welfare Changes due to Transport Projects

We assume an economy consisting of a representative individual, who has a continuous and increasing utility function that depends on the amounts chosen within a set of n consumption activities that includes all goods and services produced in this economy, $U(x_1, ..., x_n)$, where x_j represents the quantity of good or service j, with

$j = 1, \ldots, n$. This individual chooses his optimal set of consumption activities by maximizing his utility given his budget constraint. This constraint delimits all the combinations of goods and services, including leisure, that may be obtained at any given time, according to their (exogenous) market prices and individual's income, which has two components (wage and profits).

First, this individual earns income by working. Let us denote by $\bar{l}$ the maximum time endowment available for the consumer (e.g., 24 h per day, or 365 days per year), and by t_j the time required to consume each unit of good or service j.[2] Denoting by w the wage received per unit of working time, the individual's labour income is given by wl, where l represents the working time chosen by the individual, which is defined by the difference:

$$l = \bar{l} - \sum_{j=1}^{n} t_j x_j. \tag{2.1}$$

Secondly, we will assume that all firms are ultimately owned by this representative individual and that they distribute all their profits; thus, the individual's total income obtained as firm owner is given by:

$$\Pi = \sum_{j=1}^{n} \pi_j, \tag{2.2}$$

where π_j is the maximum profit obtained by firm j from producing and selling good or service j. From each firm's point of view, this profit is obtained by solving the standard maximization program:

$$\pi_j = \max_{l_j} p_j x_j^s - w l_j = p_j f_j(l_j) - w l_j, \tag{2.3}$$

where p_j is the market price of good or service j, and l_j represents the amount of labour (the only input in this model) used by firm j to produce x_j^s through the production function $f_j(l_j)$. If all the required equilibrium properties hold, the first-order condition of this problem is given by:

$$\frac{\partial \pi_j}{\partial l_j} = p_j \frac{\mathrm{d} f_j(l_j^*)}{\mathrm{d} l_j} - w = 0, \tag{2.4}$$

---

[2] Everyday life activities are time-consuming, and this resource should be explicitly included in the analysis because individuals make their travel decisions both in terms of market prices and the opportunity cost of the travel time. This follows Becker (1965) or DeSerpa (1971) but also many contributions in transport economics literature (see Jara-Díaz et al., 2016, for example).

and it allows us to obtain as a solution $\pi_j = p_j f_j\left(l_j^*\right) - wl_j^*$. Note that, in this equilibrium, the sum of all labour inputs used by firms must be equal to the working time offered by the representative individual, i.e., $\sum_{j=1}^{n} l_j^* = l$.

We can now use these results to finally define the individual's budget constraint, which is given by:

$$\sum_{j=1}^{n} p_j x_j \leq \Pi + wl, \tag{2.5}$$

which can be also rewritten as:

$$\sum_{j=1}^{n} p_j x_j \leq \Pi + w\left(\bar{l} - \sum_{j=1}^{n} t_j x_j\right),$$

that is equal to:

$$\sum_{j=1}^{n} g_j x_j \leq \Pi + w\bar{l}, \tag{2.6}$$

where $g_j = p_j + wt_j$ represents the generalized price of good or service j. For example, in the case of air transport, g includes the monetary price paid (e.g., the airline fare, airport charges, etc.) and the users' time cost (access and egress time, waiting time and flight time).[3]

Since expressions (2.5) and (2.6) are equivalent, we can write individual's budget constraint either in terms of market prices, $p = (p_1, \ldots, p_n)$, and individual's income $y = \Pi + wl$, or in terms of the generalized prices, $g = (g_1, \ldots, g_n)$, and the potential maximum income (profits income plus the value of time endowment), named here as generalized income $y^g = \Pi + w\bar{l}$.

We can now solve the individual's decision problem. If the utility function satisfies the local non-satiation property, and the budget constraint is binding, the individual's maximization problem reduces to:

$$\max_{x_1, \ldots, x_n} U(x_1, \ldots, x_n)$$

$$\text{s.t.} \sum_{j=1}^{n} p_j x_j = \Pi + wl, \tag{2.7}$$

or, equivalently, in terms of generalized prices:

---

[3] Price and value of travel time may not be the only relevant parameters affecting consumers' travel behaviour. When the overall conditions of transport services matter (in terms of comfort, reliability, safety, etc.), some additional elements of utility should be added to the generalized price. For the sake of simplicity, we omit these elements here, as the main results are unaffected.

$$\max_{x_1,\ldots,x_n} U(x_1,\ldots,x_n)$$

$$\text{s.t. } \sum_{j=1}^{n} g_j x_j = \Pi + w\bar{l}. \tag{2.8}$$

This is the preferred expression of the problem when evaluating transport projects, since most of them can be interpreted as changes in generalized prices (either due to changes in market prices and/or in travel time). Note that if a transport project reduces travel time, the individual will have more time to work (or for leisure), which in turn may lead to the production of additional goods. The project costs are measured in terms of the net monetary value of the goods that the individual has to give up to implement such a project.

The opportunity cost of travel time is the wage rate (w) in our model. This is a simplifying assumption that does not affect the main results.[4] In practice, determining the value of time often becomes an empirical question since for some individuals (those who are willing to work, but unable to find a job), the wage rate could overestimate the true opportunity cost of leisure, whereas for others, the wage rate underestimates their non-working time (when other non-monetary benefits are associated with the job). In practice, the value of travel time is usually denoted by vt_j (and not just wt_j, as assumed for simplicity in our model).[5]

The corresponding Lagrange function used to solve problem (2.8) is then given by:

$$L = U(x_1,\ldots,x_n) - \lambda \left(\sum_{j=1}^{n} g_j x_j - \Pi - w\bar{l} \right), \tag{2.9}$$

which can be also rewritten as:

$$L = U(x_1,\ldots,x_n) - \lambda \left(\sum_{j=1}^{n} g_j x_j - \sum_{j=1}^{n} p_j f_j(l_j^*) - w \sum_{j=1}^{n} t_j x_j \right). \tag{2.10}$$

---

[4] See Hensher (2011) for an overview of the major theoretical and empirical issues concerning the value of travel time savings.

[5] There are several reasons why the value of time may empirically differ from the wage rate. This is the case when both work and travel affect utility directly (and not only the budget constraint, as in our model), or when working time is unaltered by travel time savings. In those situations, the value of time of each individual depends on the sort of travel they undertake, i.e., the time at which the journey is made, the characteristics of the journey (congested, repetitive, or free flow), the journey purpose (commuting or leisure), the journey length, the mode of transport, or the size of the time saving (see Mackie et al., 2001, for further details).

First-order conditions are given by:

$$\frac{\partial L}{\partial x_j} = \frac{\partial U(x^*)}{\partial x_j} - \lambda\left(g_j - wt_j\right) = 0,$$

$$\frac{\partial L}{\partial \lambda} = \sum_{j=1}^{n} g_j x_j^* - \Pi - w\bar{l} = 0, \tag{2.11}$$

with $j = 1, \ldots, n$ and $x^* = \left(x_1^*, \ldots, x_n^*\right)$.

The solution of the above maximization program yields the Marshallian demand function for each good or service j, given by $x_j^* = x_j(g, y^g)$, with $g = (g_1, \ldots, g_n)$ representing the vector of all generalized prices, and the generalized income $y^g = \Pi + w\bar{l}$, which is defined by the sum of profits income and the value of individual's time endowment.

When the individual is maximizing his utility, the opportunity cost of one hour is the wage rate w, identified with the value of time in our model because, at the optimum, the individual is indifferent between consuming additional goods, including leisure, or working more (and giving up the corresponding units of time). Hence, the hourly wage w, is the opportunity cost of time disregarding its final use (either leisure or consumption). This is the key idea for the measurement of direct benefits of transport improvements: reducing the time spent on transport activities increases the time available for consumption of other goods, or for working. These benefits imply an opportunity cost, measured in terms of the monetary value of the other goods that the individual gives up when implementing the project.[6]

Figure 2.1 illustrates the relationship between the inverse demand function of a transport service in terms of the market price or the generalized price. As explained above, in our model, the demand function is the solution of the representative consumer's maximization program. In general, the market demand is the horizontal sum of the demand of all individuals' in the society. Thus, if x denotes the total number of trips demanded in the market, the (negative) relationship between the number of trips and the generalized price of transport services can also be interpreted as users' WTP for them, in terms of the market prices and the value of travel time.

By substituting all these demands in the (direct) utility function, we obtain the individual's indirect utility function, defined as:

$$U\left(x_1^*, \ldots, x_n^*\right) = V(g, y^g), \tag{2.12}$$

---

[6] Once the spatial nature of transport activities is included in the model, the explicit treatment of changes in proximity and location could yield potential increases of productivity and the WEBs. Thus, time savings (as measured in our model) would underestimate the social benefits of transport projects (see Sect. 2.3.3).

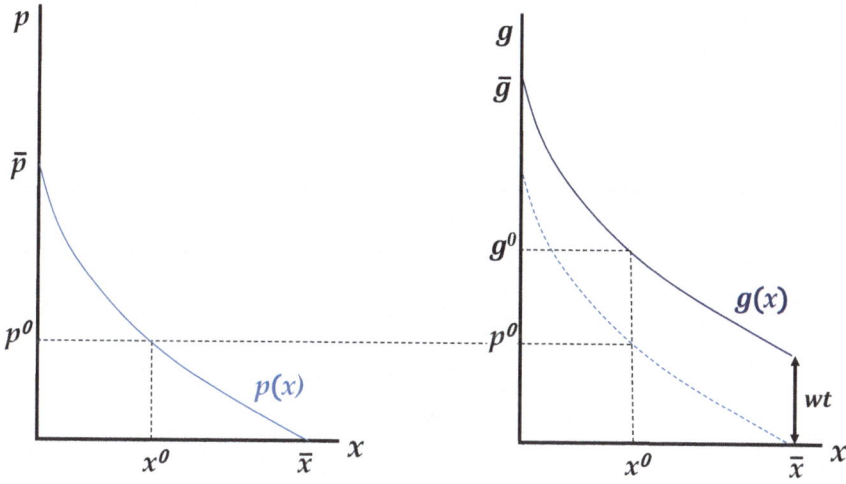

Fig. 2.1 Transport demand function: market price versus generalized price

which gives the individual's maximal attainable utility when faced with a vector g of generalized prices and individual's generalized income y^g. This utility function is called *indirect* because individuals usually think about their preferences in terms of what they consume rather than in terms of prices and income.

In addition, note that by replacing the Marshallian demands into the Lagrange function and considering first-order conditions, we have that, in equilibrium:

$$L^* = V(g, y^g) - \lambda \left(\sum_{j=1}^{n} g_j x_j^* - \Pi - w\bar{l} \right) = V(g, y^g), \qquad (2.13)$$

and therefore:

$$\frac{\partial L^*}{\partial y^g} = \lambda = V_y = \frac{\partial V(g, y^g)}{\partial y^g}, \qquad (2.14)$$

showing that the Lagrange multiplier can be interpreted as the individual's marginal utility of generalized income (V_y).

2.2.2 Measuring the Economic Effects of Transport Projects

Let us now analyse the effects of a transport project, defined as an exogenous intervention that increases the number of trips, either via investments (e.g., an increase in capacity) or other policies (such as more efficient pricing, better management practices, etc.), and/or reduces the generalized price. In our representative individual world, the change in social welfare, dW, is just given by the change in the individual's utility: dW = dU and, thus, considering the direct utility function evaluated at the initial equilibrium, we can write:

$$\mathrm{d}W = \mathrm{d}U = \sum_{j=1}^{n} \frac{\partial U(x^*)}{\partial x_j} \mathrm{d}x_j. \tag{2.15}$$

Then, substituting the first-order condition of the individual's maximization program given by (2.11) into expression (2.15), we obtain:

$$\frac{\mathrm{d}W}{V_y} = \sum_{j=1}^{n}(g_j - wt_j)\mathrm{d}x_j = \sum_{j=1}^{n} p_j \mathrm{d}x_j. \tag{2.16}$$

According to this expression, the change in social welfare resulting from a transport project that implies a marginal change in the number of trips is equal to the difference between the individual's generalized WTP for those additional trips and the value of its travel time, i.e., the market price. Note that, if the transport project has a cost, some dx_j are negative, representing the monetary value of production and consumption of other goods, including time, that the individual must give up for the project to be implemented.

Alternatively, if we use the indirect utility function, we get:

$$\mathrm{d}W = \mathrm{d}V = \sum_{j=1}^{n} \frac{\partial V}{\partial g_j} \mathrm{d}g_j + V_y \mathrm{d}y^g. \tag{2.17}$$

Applying the envelope theorem, we obtain:

$$\frac{\partial V}{\partial g_j} = -\lambda x_j = -V_y x_j, \tag{2.18}$$

which can be replaced into expression (2.17) to finally obtain a useable expression that allows us to evaluate the effects of transport projects:

$$\frac{\mathrm{d}W}{V_y} = -\sum_{j=1}^{n} x_j \mathrm{d}g_j + \mathrm{d}y^g. \tag{2.19}$$

Leaving the assumption of a representative individual, the change in social welfare is given by the sum of the change in each individual's utility, weighted by the social marginal utility of each individual. The value of the social marginal utility of income can be assumed to be equal to one, only if income distribution is optimal, or society has at its disposal means for unlimited and costless redistributions, and therefore, monetary gains and losses can be aggregated across individuals in order to determine whether the project is socially worthy. Nevertheless, redistribution is not costless since, for example, it might affect incentives in a negative way. In this case, the actual income distribution may not be far from the constrained optimal one. This means that the actual situation represents a kind of constrained optimum and possibly we can just sum gains and losses across individuals. This is also sufficient if relative prices are left more or less unchanged (see Johansson & Kriström, 2016, for further details on aggregation problems).

A Price Reduction Transport Project

Now consider that the change in the generalized price of good or service j which we interpret as a transport project is only due to a change in the market price p_j, while the required (travel) time t_j remains constant, i.e., $\mathrm{d}g_j = \mathrm{d}p_j$. In this case, we have:

$$\mathrm{d}y^g = \mathrm{d}(\Pi + w\bar{l}) = \sum_{j=1}^n \frac{\partial \pi_j}{\partial p_j}\mathrm{d}p_j = \sum_{j=1}^n x_j^s \mathrm{d}p_j. \tag{2.20}$$

By substituting this result into expression (2.19), and assuming that all product markets clear, $x_j = x_j^s$:

$$\frac{\mathrm{d}W}{V_y} = -\sum_{j=1}^n x_j \mathrm{d}p_j + \sum_{j=1}^n x_j^s \mathrm{d}p_j = 0, \tag{2.21}$$

i.e., a marginal variation in the generalized price of good or service j due to a change in the market price p_j (with t_j constant) does not produce any effect on welfare. The reason is that, if all product and labour markets clear, a change in the market price without any time saving is just a transfer between consumers and producers. Moreover, there are no other additional welfare effects to be considered in the rest of the economy.

A Time-Saving Transport Project

Alternatively, consider now that the change in the generalized price of good or service j is due to a change in time t_j while the market price p_j remains constant, i.e., $\mathrm{d}g_j = w\mathrm{d}t_j$. In this case:

$$\mathrm{d}y^g = \mathrm{d}(\Pi + w\bar{l}) = \sum_{j=1}^n w\frac{\partial \pi_j}{\partial t_j}\mathrm{d}t_j = \sum_{j=1}^n w\left(p_j\frac{\partial f_j(l_j^*)}{\partial l_j} - w\right)\frac{\partial l_j}{\partial t_j}\mathrm{d}t_j, \tag{2.22}$$

which is zero, according to the first-order condition of the profit maximization program of firm j given by expression (2.4), i.e., $\mathrm{d}y^g = 0$. Then, by substituting this into expression (2.19), we finally obtain that:

$$\frac{\mathrm{d}W}{V_y} = -\sum_{j=1}^n x_j w\mathrm{d}t_j. \tag{2.23}$$

In other words, the increase in social welfare due to a marginal reduction in travel time is equal to the value of the time savings ($dt_j < 0$) multiplied by the number of trips benefiting from that improvement.

Expressions (2.19) and (2.23) are derived by considering marginal changes with respect to the situation without the project. When the effect of a transport project is not marginal, the change in social welfare can be directly approached as the change in consumer's utility with the project with respect to the counterfactual. In our model, this change in social welfare is, thus, given by:

$$\Delta W = \Delta V = V\left(g^1, y^{g1}\right) - V\left(g^0, y^{g0}\right), \tag{2.24}$$

where superscript 1 indicates the situation 'with the project' and superscript 0 denotes 'without the project'. Thus, the social benefit of the project is expressed as the difference in the individual's utility with and without the project.

Although this utility is not directly measurable, expression (2.24) is very useful. If the individual is asked how much money he is willing to pay to enjoy the benefits derived from the reduction in the generalized price of transport due to the project, we obtain a monetary measure of the change in his utility. This is the so-called compensating variation (CV).[7] When CV is taken from the individual's income, he is indifferent between the situation with and without the project, as expressed by:

$$V\left(g^1, y^{g1} - CV\right) = V\left(g^0, y^{g0}\right). \tag{2.25}$$

If the project implies costs, the CV does not only account for the benefits of the project but also for the negative effects on utility derived from the diversion of goods and labour from other uses (i.e., the cost of the project). Therefore, the CV represents the change in the generalized WTP due to the project benefits minus the willingness to accept for the goods and labour required by the project. The net social value of the government intervention is then:

$$\Delta W = CV = \Delta WTP - \Delta Resources. \tag{2.26}$$

Time savings, the main benefit in many transport projects, can be considered either as an increase in the WTP or a positive change in resources. This is not important although, given the position of a generalized demand curve (see Fig. 2.1), the decrease in the generalized price of transport with the project increases the number of trips, and thus a change in the WTP of this generated demand. For the existing demand, the WTP (including time) has not changed, and thus, we can consider the value of time savings as a (positive) change in resources.

---

[7] Alternatively, if the individual is asked how much money he is willing to accept to waive the potential benefits derived from the reduction in the generalized price of transport due to the project, we also obtain a monetary measure of the change in his utility. This is the so-called equivalent variation (EV). When EV is given to the individual as an additional income, he is indifferent between the situation with and without the project, as expressed by: $V\left(g^1, y^{g1}\right) = V\left(g^0, y^{g0} + EV\right)$. If income effects are zero, CV and EV coincide.

Suppose the representative individual is asked for his WTP for the transport project disregarding any effects on his profits income. Then, the maximum WTP, CV, as defined in expression (2.25), and the new partial one, denoted by CV^P are given by:

$$CV = CV^P + \Delta PS, \tag{2.27}$$

where ΔPS represents the change in firms' profits due to the transport project. If income effects are not significant, CV^P can be approximated through the change in consumer's surplus (CS),[8] and then:

$$\Delta W = CV \approx \Delta CS + \Delta PS, \tag{2.28}$$

i.e., social welfare changes can be approximated through the sum of the changes in the surpluses of consumers and producers.

2.3 Practical Rules for Cost–Benefit Analysis

So far, we have described the foundations of the two main approaches to measure the net benefits of transport projects: adding the changes in WTP and the use of resources or adding the changes in consumers' and producers' surpluses. However, expressions (2.24) to (2.28) can be generalized to include other roles of the individual in the society. A practical disaggregation is to consider three owners of production factors: first, the 'owners of capital' (O), generally called producers, who have a variety of equipment, infrastructure and facilities where goods and services are produced; second, the 'owners of labour' (L) including for simplicity employees of different skills and productivity levels, and the 'landowners' (R).

The fixed factor 'land' is restricted here to soil for agriculture or land for residential or productive uses. We differentiate the landowners from the common property of natural and environmental resources (also called 'natural capital'). Natural and environmental resources such as climate, water, air, flora and fauna and landscapes, which may be affected by projects, are included in 'rest of society'(E). Adding consumers (C) and taxpayers (G), six roles for the representative individual are identified for evaluation purposes (de Rus, 2021a). The rationale for this disaggregation is both for minimizing the risk of double counting and for equity considerations.[9]

---

[8] The relative error of using the change in consumer's surplus instead of CV^P is low if the elasticity of demand with respect to income, or the proportion of the change in consumer's surplus with respect to income, is small enough (Willig, 1976).

[9] Important practical issues arise if capital or land is under foreign ownership (see Johansson & de Rus, 2019). Our analysis assumes a closed economy.

Following Johansson (1993), the individual's indirect utility function is now given by $V(p, t, w, \Pi, \tau, z)$, where $p = (p_1, \ldots, p_n)$ is the vector of market prices, $t = (t_1, \ldots, t_n)$ is the vector of the time required for consuming each good or service, w is the wage, Π is firms' profits, τ is a lump-sum tax, and z represents a set of natural resources.

In this setup, the change in social welfare due to a transport project (which implies a reduction in transport generalized price) is given by:

$$\Delta W = \Delta V = V\left(p^1, t^1, w^1, \Pi^1, \tau^1, z^1\right) - V\left(p^0, t^0, w^0, \Pi^0, \tau^0, z^0\right), \quad (2.29)$$

and using the concept of CV, we have that:

$$V\left(p^1, t^1, w^1, \Pi^1, \tau^1, z^1 - \text{CV}\right) = V\left(p^0, t^0, w^0, \Pi^0, \tau^0, z^0\right), \quad (2.30)$$

with:

$$\text{CV} = \text{CV}^P + \Delta\text{OS} + \Delta\text{LS} + \Delta\text{RS} + \Delta\text{GS} + \Delta\text{ES}, \quad (2.31)$$

where CV^P can be approximated by changes in consumers' surplus; ΔOS is the change in firm's revenues minus variable costs; ΔLS refers to the change in workers' surplus; and ΔRS is the landowners' surplus, equal to the wage and land income, respectively, minus the minimum payment they are willing to accept for the use of the factor, i.e., its *private* opportunity cost; ΔGS is the change in taxpayers' surplus, equals tax revenues minus public expenditure; and ΔES is the change in the surplus of the rest of society, i.e., the value of the externality minus the compensations received (if any).

Finally, adding the changes in surpluses, the income transfers net out, and it is easy to show that the result is again equal to the change in WTP minus (plus) the value of the diverted (saved) goods and labour from other uses and the negative (or positive) external effects:[10]

$$\Delta W = \Delta\text{CS} + \Delta\text{OS} + \Delta\text{LS} + \Delta\text{RS} + \Delta\text{GS} + \Delta\text{ES} = \Delta\text{WTP} - \Delta\text{Resources}. \quad (2.32)$$

2.3.1 Guidelines for the Measurement of the Direct Effects

For the sake of simplicity, consider a transport project without investment costs. The change in welfare with the project is measured with $g(x)$, the market inverse derived demand function for a transport activity in terms of its generalized price. The initial

---

[10] Notice that an external effect is a change in resources.

equilibrium is (g^0, x^0), and marginal operating costs are constant and equal to c. With the project, the generalized price of transport is reduced to g^1, and the number of trips increases to x^1.

Once the benefits and costs of the project are identified, the practitioner has to choose one of the available alternative approaches for the measurement of those costs and benefits. A clear understanding of the chosen method avoids common errors that may lead to the overestimation or underestimation of the net benefit.[11]

Adding the change in surpluses (expression 2.32) is straightforward and provides more information, but it is difficult to be applied in practice given the data usually available and the difficulty of the ex-ante identification of the final beneficiaries. The alternative consists in identifying and measuring the changes in WTP and resources, and though it seems, at first sight, easier, it has some pitfalls associated with its use.

CBA guidelines usually present a single graph for the transport mode directly affected by the project, showing the change in the generalized price and quantities, decomposed in benefits for the existing demand and for deviated and generated demand. The analysis may look multimodal when deviated demand is included but in practice is prone to errors when price deviates from marginal cost in other modes or activities where generated demand comes from.

One easy way to proceed when in other affected parts of the transport system prices are not equal to marginal social costs is to consider a unique corridor where all transport alternatives operate and then add the changes in the surpluses of all the affected agents (expression 2.32) and modes. Alternatively, using the change in WTP and resources, the corridor assumption is very helpful in the presence of taxes or market power. In a corridor between A and B, there is no change in WTP (no change in quality is assumed for simplicity) because the WTP depends on the purpose of the trip and it is not necessarily affected by the change of route, time, or mode under the assumption of constant quality (transport is commonly a derived demand).

Therefore, the change from g^0 to g^1 only translates into the use of new resources absorbed by the project and the saving (substitutes) and consumption (complements) of additional resources in the rest of the modes affected within the corridor. This is even so in the case of generated demand coming from other consumption activities (and, hence, these activities need to be included in our corridor). In the initial equilibrium, the marginal unit of generated demand was indifferent between travelling and the other consumption activities, so the treatment is identical to any transport mode. As the effects in many secondary markets can be of different signs and many of them are simple relocation, the sensible way to proceed is to follow the corridor (or multimodal) analysis including some reasonable assumptions based on the best information available on the source of the generated demand coming outside the transport market.

---

[11] There are all sorts of measurement/prediction errors, which apply to both methods (Mackie & Preston, 1998).

Hence, we can summarize the two alternative approaches: the first one, adding the surpluses of all the agents involved in all transport modes, and some other economic activities affected by the project. Nevertheless, when the price is equal to marginal cost in the rest of the economy, it is correct to concentrate only on the mode directly affected by the change in the generalized price, disregarding intermodal effects.

When the practitioner decides to follow the change in WTP and resources, ignoring transfers, there are two options: (i) the conventional single graph analysis, common to many CBA guidelines, where the WTP is constant for existing users (assuming quality to be constant) but there is an increase in WTP of deviated and generated demand. In this case, we have to add any distortion (e.g., loss of profits or taxes) in the other modes and economic activities affected by the change in the primary market. In this case, it is incorrect to include the change in resources used or saved in the secondary markets. Moreover, if the practitioner ignores the effects due to taxes and market power in the other modes and economic activities, there is a measurement error. (ii) The corridor analysis, where the change in WTP is limited to any change in quality or safety and only changes in resources are accounted for. In this case, the practitioner should include any change in resources used or saved in the original transport mode and any other mode or economic activity included in the corridor.

Let us formally prove the equivalence of the different approaches. Consider a market with n modes of transport or activities, where a transport project reduces the generalized price of mode i $(g_i^1 < g_i^0)$ because of a reduction in travel time $(t_i^1 < t_i^0)$. Travel time includes access, waiting, in-vehicle, and egress time. Note that, although there is a reduction in generalized price, it is possible to charge a higher price $(p_i^1 > p_i^0)$, though it must be lower than the reduction in the value of the time component. Let us denote by v_i the value of time of users initially travelling in mode i, by c_i the constant marginal operating cost of mode i (assumed to be different with the project), and by τ_i the value of an *ad valorem* tax applied to mode i. Notice that, since there is an *ad valorem* tax, the price charged by producers (p_i^-) does not coincide with the price paid by users (p_i), where $p_i = p_i^-(1 + \tau_i)$.

We assume that the value of time for users initially choosing an alternative mode or activity j (v_j) is different than the value of time for users initially travelling in mode i (v_i);[12] there are *ad valorem* taxes in all the alternatives (τ_j), so the price charged by producers (p_j^-) does not coincide with the price paid by users (p_j), where $p_j = p_j^-(1 + \tau_j)$; and marginal operating cost is constant in each alternative mode or activity (c_j), with $j = 1, \ldots, n$ and $j \neq i$. We also assume that alternative j generates an externality equal to E per passenger-trip. Finally, income effects are not significant.

According to expression (2.32), the change in social welfare is the sum of the changes in surpluses of all the agents affected in all transport modes and in other economic activities, affected by the project, which can be easily calculated using the standard assumption of a linear approximation between the initial and the final

---

[12] Notice that in this section, we use the parameter v to denote the value of time since, as already mentioned, it may empirically differ from the wage rate.

generalized prices (the so-called 'rule of a half').[13] We distinguish between existing demand (users already travelling in mode i), deviated demand (users changing from an alternative mode with the project), and generated demand (coming from other consumption activities). We follow the same procedure for deviated and generated demand since the former comes from other modes and the latter comes from other activities. Using the superscripts e and d to denote changes due to existing demand, and deviated and generated demand from mode or activity j, respectively, the transport project implies a change in social welfare given by:

$$\Delta W = \Delta W^e + \sum_{\substack{j=1 \\ j \neq i}}^{n} \Delta W_j^d. \tag{2.33}$$

For existing demand (x_i^0), the benefits of the project come from the change in consumer's surplus of existing users, change in firm's revenues minus variable costs in this traffic, and change in tax revenues (because of the price increase):[14]

$$\Delta CS^e = \left(g_i^0 - g_i^1\right)x_i^0 = \left(p_i^0 + v_i t_i^0\right)x_i^0 - \left(p_i^1 + v_i t_i^1\right)x_i^0, \tag{2.34}$$

$$\Delta OS^e = \left(p_i^{1-} - p_i^{0-}\right)x_i^0 - \left(c_i^1 - c_i^0\right)x_i^0, \tag{2.35}$$

$$\Delta GS^e = \tau_i\left(p_i^{1-} - p_i^{0-}\right)x_i^0, \tag{2.36}$$

$$\Delta ES^e = 0. \tag{2.37}$$

Hence, the change in social welfare due to the existing demand is given by:

$$\Delta W^e = \Delta CS^e + \Delta OS^e + \Delta GS^e + \Delta ES^e = v_i(t_i^0 - t_i^1)x_i^0 - (c_i^1 - c_i^0)x_i^0. \tag{2.38}$$

In the case of deviated demand from mode or activity j, $g_j^0 = p_j + v_j t_j^0$ denotes the generalized price for the user indifferent between mode or activity j and mode i without the project, where t_j^0 denotes the travel time of such an indifferent user. Notice that in the initial equilibrium, g_j^0 has to be equal to $g_i^{0d} = p_i^0 + v_j t_i^0$ (i.e., without the project, the indifferent user has the same generalized price in mode or alternative j, g_j^0, and in conventional train, g_i^{0d}). All those users with generalized price in mode or activity j higher than the generalized price of the indifferent user $g_j^0 = g_i^{0d}$ had chosen mode i instead of this alternative. On the contrary, all those users with generalized price in mode or activity j lower than the generalized price

---

[13] See Harberger (1965), Neuberger (1971) and Small (1999).

[14] We assume no change in workers' surplus nor landowners' surplus.

of the indifferent user $g_j^0 = g_i^{0d}$ had chosen mode or alternative j instead of mode i. Once the project is implemented, the generalized price in mode i is reduced to $g_i^{1d} = p_i^1 + v_j t_i^1$ and, due to this reduction, some users that preferred mode or activity j before the project now prefer mode i. Thus, x_j^d represents the deviated demand from mode or activity j to mode i, and total demand with the project (x_i^1) is equal to $x_i^0 + \sum_{\substack{j=1 \\ j \neq i}}^n x_j^d$. Now, there is a new indifferent consumer, and his generalized price in the alternative is $g_j^1 = p_j + v_j t_j^1$, where t_j^1 denotes the travel time of this new indifferent consumer once the project has been implemented. Notice that t_j^1 is different than t_j^0 since, for example, consumers have different access or egress time. Finally, similarly to the former indifferent user, in the final equilibrium, g_j^1 has to be equal to $g_i^{1d} = p_i^1 + v_j t_i^1$ for the new one.

Adding the change in surpluses for deviated demand, the benefits of the project come from the change in consumers' surplus of the deviated users from mode or activity j (linear approximation), change in firm's revenues minus variable costs (firms i and j), change in collected taxes, and change in the surplus of the rest of society (equal to the value of the externality as, for simplicity, compensations are assumed to be zero).

$$\Delta CS_j^d = \frac{1}{2}(g_j^0 - g_i^{1d})x_j^d = \frac{1}{2}[(p_j + v_j t_j^0) - (p_i^1 + v_j t_i^1)]x_j^d, \tag{2.39}$$

$$\Delta OS_j^d = \left(p_i^{1-} - p_j^-\right)x_j^d - (c_i^1 - c_j)x_j^d, \tag{2.40}$$

$$\Delta GS_j^d = \tau_i p_i^{1-} x_j^d - \tau_j p_j^- x_j^d, \tag{2.41}$$

$$\Delta ES_j^d = -Ex_j^d. \tag{2.42}$$

Hence, the change in social welfare due to the deviated demand from mode or activity j is:

$$\begin{aligned}
\Delta W_j^d &= \Delta CS_j^d + \Delta OS_j^d + \Delta GS_j^d + \Delta ES_j^d \\
&= \frac{1}{2}v_j(t_j^0 - t_i^1)x_j^d + \frac{1}{2}(p_i^1 - p_j)x_j^d - (c_i^1 - c_j)x_j^d - Ex_j^d.
\end{aligned} \tag{2.43}$$

Finally, following the change in surpluses approach, the change in social welfare for the whole demand, adding (2.38) and (2.43), is:

$$\begin{aligned}
\Delta W &= \Delta W^e + \sum_{\substack{j=1 \\ j \neq i}}^n \Delta W_j^d \\
&= v_i(t_i^0 - t_i^1)x_i^0 - (c_i^1 - c_i^0)x_i^0
\end{aligned}$$

$$+ \sum_{\substack{j=1 \\ j \neq i}}^{n} \left[\frac{1}{2} v_j \left(t_j^0 - t_i^1 \right) x_j^d + \frac{1}{2} (p_i^1 - p_j) x_j^d - (c_i^1 - c_j) x_j^d - E x_j^d \right]. \quad (2.44)$$

Adding the changes in WTP and resources following the unimodal or single graph analysis, the change in social welfare is equal to the change in WTP and the change in resources. First, for the existing demand, the change in WTP (ΔWTP^e) is zero, and the change in resources (ΔResources^e) is equal to the value of the time invested (saved in our project because $t_i^1 < t_i^0$) and the change in operating cost of existing trips:

$$\Delta W^e = \Delta \text{WTP}^e - \Delta \text{Resources}^e = -v_i \left(t_i^1 - t_i^0 \right) x_i^0 - (c_i^1 - c_i^0) x_i^0. \quad (2.45)$$

It is immediate to check that Eqs. (2.38) and (2.45) coincide.

Second, the change in WTP and the change in resources due to the deviated demand from mode or activity j are equal to:

$$\Delta \text{WTP}_j^d - \Delta \text{Resources}_j^d = \frac{1}{2} \left(g_j^0 + g_i^{1d} \right) x_j^d - v_j t_i^1 x_j^d - c_i^1 x_j^d - E x_j^d. \quad (2.46)$$

Expression (2.46) shows the difference between the increase in the users' WTP for the new trips (ΔWTP$_j^d$, deviated from mode or activity j) and the resources required to obtain those benefits (ΔResources$_j^d$), i.e., the value of the time spent on the new trips and the operating cost of the new trips. Notice that expression (2.46) does not coincide with the change in social welfare (ΔW_j^d) given by expression (2.43). There is a measurement error because the practitioner is ignoring the effects due to taxes in the other modes or economic activities. Adding such effects, the change in social welfare due to the deviated demand is given by:

$$\begin{aligned}
\Delta W_j^d &= \frac{1}{2} \left(g_j^0 + g_i^{1d} \right) x_j^d - v_j t_i^1 x_j^d - c_i^1 x_j^d - E x_j^d - \tau_j p_j^- x_j^d \\
&= \frac{1}{2} \left(g_j^0 + g_i^{1d} \right) x_j^d - v_j t_i^1 x_j^d - c_i^1 x_j^d - E x_j^d - (p_j - c_j) x_j^d \\
&= \frac{1}{2} \left[(p_j + v_j t_j^0) + (p_i^1 + v_j t_i^1) \right] x_j^d - v_j t_i^1 x_j^d - c_i^1 x_j^d - E x_j^d - p_j x_j^d + c_j x_j^d \\
&= \frac{1}{2} v_j \left(t_j^0 - t_i^1 \right) x_j^d + \frac{1}{2} (p_i^1 - p_j) x_j^d - (c_i^1 - c_j) x_j^d - E x_j^d, \quad (2.47)
\end{aligned}$$

which is equal to expression (2.43).

At this point, we would like to highlight the importance of having a theoretical model as a reference for practical CBA, avoiding the mechanical application of rules of thumb from different sources that imply double counting and measurement errors.

Finally, the change in social welfare for the whole demand, adding (2.45) and (2.47), is:

$$
\begin{aligned}
\Delta W &= \Delta W^e + \sum_{\substack{j=1 \\ j \neq i}}^{n} \Delta W_j^d \\
&= -v_i (t_i^1 - t_i^0) x_i^0 - (c_i^1 - c_i^0) x_i^0 \\
&\quad + \sum_{\substack{j=1 \\ j \neq i}}^{n} \left[\frac{1}{2} v_j (t_j^0 - t_i^1) x_j^d + \frac{1}{2} (p_i^1 - p_j) x_j^d - (c_i^1 - c_j) x_j^d - E x_j^d \right], \quad (2.48)
\end{aligned}
$$

which is equal to expression (2.44).

Alternatively, we may add the changes in WTP and resources following the multi-modal or the corridor analysis. The change in social welfare is equal to the saved operating cost plus time savings. No change in WTP occurs within the corridor as, by assumption, the modal change does not affect the quality of travel. For existing demand, the change in social welfare following the multimodal or the corridor analysis is given by:

$$
\Delta W^e = \Delta \text{Resources}^e = (c_i^0 - c_i^1) x_i^0 + v_i (t_i^0 - t_i^1) x_i^0, \quad (2.49)
$$

that is equal to (2.38) and (2.45).

For deviated demand, to calculate the change in social surplus, we have to take into account the cost and time saved in the alternative mode or activity j, and the cost and time spent in mode i. In other words, we must compute the cost and time saved by deviated demand shifting from alternative mode or activity j to mode i. Finally, we have to consider the externality.

Regarding the time saved by each consumer shifting from alternative j to mode i, it should be highlighted that time savings are not the same for everyone who deviated from the alternative mode. Time savings for the indifferent consumer without the project are the highest and equal to $v_j (t_j^0 - t_i^1)$, while time savings for the new indifferent consumer with the project are the lowest and equal to $v_j (t_j^1 - t_i^1)$. Time savings are given by $\frac{1}{2} v_j \left[(t_j^0 - t_i^1) + (t_j^1 - t_i^1) \right] x_j^d$. Moreover, time savings could be also computed as:

$$
\begin{aligned}
&\frac{1}{2} \left[((g_j^0 - p_j) - (g_i^{1d} - p_i^1)) + ((g_j^1 - p_j) - (g_i^{1d} - p_i^1)) \right] x_j^d \\
&= \frac{1}{2} v_j \left[(t_j^0 - t_i^1) + (t_j^1 - t_i^1) \right] x_j^d. \quad (2.50)
\end{aligned}
$$

Thus, adding cost saving and externalities, the change in social welfare is:

$$\Delta W_j^d = \Delta \text{Resources}_j^d$$
$$= \frac{1}{2}v_j(t_j^0 - t_i^1)x_j^d + \frac{1}{2}(t_j^1 - t_i^1)x_j^d + (c_j^0 - c_i^1)x_j^d - Ex_j^d. \qquad (2.51)$$

Recall that for the new indifferent user the generalized price is g_j^1 and equal to g_i^{1d}. Therefore, we can rewrite expression (2.50) as:[15]

$$\frac{1}{2}[((g_j^0 - p_j) - (g_i^{1d} - p_i^1)) + ((g_i^{1d} - p_j) - (g_i^{1d} - p_i^1))]x_j^d$$
$$= \frac{1}{2}v_j(t_j^0 - t_i^1)x_j^d + \frac{1}{2}(p_i^1 - p_j)x_j^d. \qquad (2.52)$$

Thus, adding cost saving and externalities, the change in social welfare given by expression (2.51) could be rewritten as:[16]

$$\Delta W_j^d = \Delta \text{Resources}_j^d$$
$$= \frac{1}{2}v_j(t_j^0 - t_i^1)x_j^d + \frac{1}{2}(p_i^1 - p_j)x_j^d + (c_j^0 - c_i^1)x_j^d - Ex_j^d, \qquad (2.53)$$

that is equal to (2.43) and (2.47).

Finally, the change in social welfare for the whole demand, adding (2.49) and (2.53), is:

$$\Delta W = \Delta W^e + \sum_{\substack{j=1 \\ j\neq i}}^{n} \Delta W_j^d$$
$$= (c_i^0 - c_i^1)x_i^0 + v_i(t_i^0 - t_i^1)x_i^0$$
$$+ \sum_{\substack{j=1 \\ j\neq i}}^{n}\left[\frac{1}{2}v_j(t_j^0 - t_i^1)x_j^d + \frac{1}{2}(p_i^1 - p_j)x_j^d + (c_j^0 - c_i^1)x_j^d - Ex_j^d\right], \qquad (2.54)$$

which is equal to expression (2.44) and (2.48).

Finally, it is worth highlighting that the distinction between the surpluses of different agents in expression (2.31) and (2.32) shows the difficulty of identifying ex-ante the final beneficiaries of the transport improvement. The explicit consideration

---

[15] Notice that, since $g_j^0 = g_i^{0d}$, time savings given by expression (2.50) may be also expressed as: $\frac{1}{2}v_j(t_i^0 - t_i^1)x_j^d + \frac{1}{2}[(p_i^0 - p_j) + (p_i^1 - p_j)]x_j^d$.

[16] It is common to consider that time savings of deviated demand are given by $\frac{1}{2}v_j\left(t_j^0 - t_i^1\right)$, but this is only the case if $p_j = p_i^1$.

of a fixed factor (such as land) in the social surplus expression may help in the under-standing of one of the main sources of double counting in transport project appraisal, helping also to clarify the distribution of the social surplus. It is well known that land can capitalize most of the benefits of transport improvements. In the case of an infinitely elastic supply of homogeneous workers, the surplus of each group in expression (2.32), taxpayers' surplus excepted, would be zero, and the landowners would take the total surplus through higher land prices.

Furthermore, Collier & Venables (2018) have shown that with heterogeneity, both in labour productivity and demand for housing, workers can gain a significant part of the surplus. The implication for the economic evaluation of transport improvements is that although a project increased the land value around the locations affected by the improvement, only in some extreme cases, this increase would reflect the total benefits of the projects because a share of those benefits is captured by workers. This leads to a practical conclusion: it is easier to calculate the change in WTP and the change in resources than to know who finally get the surplus.

Thus, the conclusion that transport benefits could be measured in a competitive land market when this market is not affected by bubbles and speculation or any other exogenous factors only holds under some restrictive conditions. What is true, in any case, is that the practitioner should be very careful avoiding the combined use of the three possible approaches: change in surpluses, change in WTP and resources, or the increase in land prices.

2.3.2 Rules for the Use of Shadow Prices

The social benefits achieved through the reduction of the generalized price of trans-port are not free. These benefits have an opportunity cost that is measured by the value of resources diverted from other uses to the project. This section deals with the inputs needed for the transport project and how to value them.[17]

The effects of a transport project on social welfare can be expressed as the maximum income the affected individuals are willing to pay to enjoy the corre-sponding benefits, net of the project costs. This is the value of the sum of the CVs as in the left-hand side of expression (2.32) for all the individuals of the society, which is net of project costs. The aggregation of the CV is then the sum of the individuals' WTP for the benefits of the project (positive sign) and the willingness to accept for giving up other goods to achieve those benefits (negative sign). This net value is approached with the right-hand side of expression (2.32).

Thus, the social opportunity cost of the project (C_j) can be defined as the value of all the goods the society has to give up when those resources are deviated from other uses to implement the project, i.e., to enjoy the utility of good j (e.g., a faster transport service), as formally represented by:

---

[17] See Johansson (1993) and de Rus (2021a). In particular, the section deals with inputs that can be purchased in markets. Non-market resources are not discussed here.

$$C_j = \sum_{k=1}^{s} p_k dx_k, \tag{2.55}$$

an expression derived from the model in Sect. 2.2.1, with $s \leq n$ goods or services, and where the only input, labour, is fully utilized to produce and consume goods and assuming that market prices reflect the value of the goods deviated to the project.

The problem is that this expression is not very useful for computing the net social benefit of a project. The practitioner generally does not know which goods (schools, housing, leisure facilities, etc.) the society gives up to achieve the benefits of the project under assessment. However, there is a way to circumvent this problem. To do so, we now consider that the production of any of the goods in this expression, x_k, requires labour, and the corresponding production function is then given by:

$$x_k = f_k(l_k), \tag{2.56}$$

whose differential shows that any output variation depends on the change in the quantity of the input multiplied by its marginal productivity:

$$dx_k = \frac{d f_k(l_k)}{d l_k} dl_k. \tag{2.57}$$

Replacing expression (2.57) in (2.55) and recalling that any profit maximizing firm uses additional units of input until its market price equals the value of its marginal productivity, $w = p_k \frac{d f_k(l_k)}{d l_k}$ (see expression 2.4 in Sect. 2.2.1), the cost of the project can be expressed as:

$$C_j = \sum_{k=1}^{s} w dl_k. \tag{2.58}$$

The cost of the project initially expressed in (2.55) as the social value of the diverted goods, to get the good provided by the project appears now in (2.58) as the quantity of labour required to produce those goods, dl_k, multiplied by the price of labour, w.

In practice, the validity and usefulness of expression (2.58) for identifying and calculating the costs of a project are conditioned by three underlying assumptions. Firstly, all the changes in input markets (in our case, labour market) are marginal; secondly, input markets are perfectly competitive, without distortions (such as indirect or income taxes); and thirdly, all the resources are fully utilized. Nevertheless, once these assumptions are abandoned to deal with more realistic project assessment situations (that include, among others, the presence of subsidies or taxes, or the use of unemployed labour in the project), expression (2.58) is no longer valid to calculate the opportunity cost of the project. This is what shadow pricing is about: adjusting market prices to reflect the true opportunity cost; and beware that this adjustment only applies in the change in WTP and resources approach.

Recall that in our model, we are considering that there is only one input: labour. Although in actual projects, there may be more inputs (a transport project typically requires the use of some produced goods, such as vehicles, energy, spare parts, and other materials),[18] the analysis of the shadow price of labour is virtually the same as the one applied to other inputs. Therefore, we will restrict our discussion to the shadow price of labour.

Labour is required in the design and construction of transport infrastructure, in its maintenance and operation, and in the provision of transport services using that infrastructure. The opportunity cost of labour in expression (2.58) is valued at its market price, w, but again this is only valid under several restrictive assumptions that usually do not hold in actual project assessments, particularly with unemployment. Thus, once the amount of labour required for the project is known, the next step is to identify where this input comes from. Suppose now that we refer to labour as the number of workers required for the project. In the analysis of the shadow price of labour, it is advisable to distinguish three main possible sources of the labour demanded by a project: (a) workers already employed in other productive activities; (b) voluntarily unemployed at the current wage; and (c) involuntarily unemployed, willing to work at the current wage.

We will assume that the project will have a significant effect on the demand for labour and that there is a proportional income tax, τ_w. Initially, without the project, the labour market is in equilibrium with the supply (S) and demand (D^0) determining a wage rate of w^0 and a quantity of labour of L^0. The existence of a proportional income tax (τ_w) introduces a distinction between the market supply function (S) and the opportunity cost of the labour supplier, $S(1 - \tau_w)$. The function $S(1 - \tau_w)$ shows the marginal value of leisure of the workers and the demand function is the value of the marginal productivity of labour for the firm. At the equilibrium wage rate (w^0), the value of the marginal productivity of labour for the firm is equal to the value of leisure for the marginal worker plus the income tax.

With the project, the demand for labour shifts from D^0 to D^1, the wage rate goes up to w^1, and the private demand for labour goes down until w^1 is equal to the value of the marginal productivity of labour. The increase in the wage rate has also the effect of increasing the number of workers willing to work at this higher wage rate, and the equilibrium number of workers goes up. Now, we can calculate the opportunity cost of labour. The project needs ΔL units of labour. This quantity of labour required by the project has two components: new workers (ΔL_n) that are willing to work at the new equilibrium wage, and workers already employed in the private sector (ΔL_p), who shift to the project at the higher wage w^1. The opportunity cost of previously voluntarily unemployed workers (ΔL_n) is the value of leisure forgone when they accept the new jobs. They are paid $w^1(1 - \tau_w)\Delta L_n$, and their social opportunity cost is:

---

[18] Note that the distinction between goods and inputs is somehow blurred in practice as the inputs to be purchased for the project are indeed produced inputs (i.e., goods). Nevertheless, the distinction is useful for the discussion of the shadow price of inputs when those inputs deviate from the private sector.

$$\left[\frac{1}{2}\left(w^0 + w^1\right)(1 - \tau_w)\right]\Delta L_n. \tag{2.59}$$

The opportunity cost of those already working in the private sector $\left(\Delta L_p\right)$, who shift to the project at the higher wage w^1, is also $w^1(1 - \tau_w)$. However, the social opportunity cost of these workers is higher than the former expression and equal to the lost value of their marginal productivity in the private sector when the amount of labour $\left(\Delta L_p\right)$ shifts to the project. They are paid $w^1(1 - \tau_w)\Delta L_p$, but the social opportunity cost of these workers is in principle:

$$\frac{1}{2}\left(w^0 + w^1\right)\Delta L_p. \tag{2.60}$$

This is the opportunity cost of the deviated labour when w represents the unit cost of labour for the firm. In the case of a proportional social security contribution paid by employers (α_w) plus the existence of *ad valorem* indirect taxes (e.g., value added tax) levied on the product market, the shadow price of the deviated labour has to reflect the social value lost as a consequence of displacing labour from other productive activities. This includes the tax revenues and any other charges lost in the process. The shadow price of labour is in this latter case:

$$(1 + \theta)(1 + \alpha_w)\left[\frac{1}{2}\left(w^0 + w^1\right)\Delta L_p\right]. \tag{2.61}$$

When the labour of the project is involuntarily unemployed, willing to work at the current wage, the supply has an infinite elasticity showing that the workers are willing to work at the equilibrium wage if they are hired by the firms. At the level of demand D^0, there is involuntary unemployment. The project shifts the demand for labour from D^0 to D^1. The project requires ΔL_n units of labour, and this amount is supplied without any change in the initial wage rate. We assume here the existence of unemployment benefits equal to u and a proportional income tax (τ_w) if the individual accepts the job.

It is useful to distinguish between the worker opportunity cost and the social opportunity cost. When the worker receives unemployment benefits equal to u, and there is a proportional income tax (τ_w), the worker's reservation wage is w^0 (he is not willing to work for less than this wage), so the workers' payment is equal to the value of leisure plus the unemployment benefits (u) and the income tax (τ_w) the worker must pay if he accepts the job.

The individual opportunity cost is wage net of taxes, equal to the value of leisure $[w^0(1 - \tau_w) - u]$ plus the employment benefits (u). However, the social opportunity cost does not include the unemployment benefits (which is a mere transfer) as a cost of the project because the real loss in resources when the individual is employed is simply the marginal value of leisure. The shadow price of labour is then $w^0(1 - \tau_w) - u$, and the social cost of these workers for the project is $[w^0(1 - \tau_w) - u]\Delta L_n$.[19]

Both values (private and social opportunity costs) can be used in the economic evaluation of projects. The point is to be consistent with the chosen approach. In the case of adding the change in surpluses, the private opportunity cost is what matters $[w^0(1 - \tau_w)]$, and the shadow price should be ignored, whereas the social opportunity cost $[w^0(1 - \tau_w) - u]$ must be used when the approach followed is the change in WTP and resources.

For the sake of exposition, we assume that the project's good is provided free of charge. Let us add the change in surpluses: the change in consumers' surplus is the total WTP, and the capital owners' surplus is equal to $-w^0 \Delta L_n$. Notice that there is no change in workers' surplus as they are paid their private opportunity cost. Finally, the taxpayers' surplus increases in the income tax collected $\tau_w w^0 \Delta L_n$ and the unemployment benefit payments avoided $u\Delta L_n$. Therefore, the net social surplus is equal to the change in consumers' surplus (WTP for the good provided by the project) plus the change in the capital owners' surplus $(-w^0\Delta L_n)$ plus the change in taxpayers' surplus $((\tau_w w^0 + u)\Delta L_n)$, i.e., the change in WTP minus the social cost of the project, $[w^0(1 - \tau_w) - u]\Delta L_n$.

2.3.3   Indirect Benefits and Wider Economic Benefits of Transport Projects. What to Do?

This section has no intention to review or comprehensively discuss the issue of indirect effects and WEBs. On the contrary, it only tries to warn of the risk of generalizing some empirical results, which are context-specific, as it is the case of the economies of agglomeration and other additional benefits used to justify projects with poor social value. This is also the case, for example, of including multiplier effects using computable general equilibrium models to estimate the total impact of the intervention instead of its net social effect. The impact of the project includes the multiplier effects but not the net welfare effect when these induced effects on the rest of the economy, as it is usually the case, are common to the counterfactual.

The idea of the indirect effect is straightforward and has already been discussed in the treatment of the measurement of the direct benefits and the multimodal effects. The welfare effects of transport improvements should not ignore the possibility of significant indirect effects, beyond transfers and relocation, and even the presence of WEBs. The spatial nature of transport introduces other benefits from increases in productivity through different mechanisms, such as industrial reorganization and also changes in land use. In any case, the qualifications about the calculation of the social surplus using expression (2.32) still apply.

---

[19] Assuming a shadow price of public funds equal to one.

Indirect effects and WEBs need some market distortion to play some influence in the economic evaluation of projects. The effects of transport improvements in secondary markets can be ignored if the rest of the economy is perfectly competitive. The indirect effects in transport projects go from intermodal effects to the impacts on some secondary markets. In both cases, the products of the secondary markets are complements or substitutes of the primary transport market. The treatment of these so-called indirect effects is similar for any secondary market (Harberger, 1965; Mohring, 1971).

The common practice is to ignore the indirect effects, under the assumption of perfectly competitive markets, or the existence of different effects in the economy, and the similar second-order general equilibrium effects of alternative investments. In any case, the practitioner should include the group of strongly interrelated markets in demand and supply. This is the case of intermodal effects which could be treated within an integrated primary transport market (see Sect. 2.3.1), or if considered as separated markets, included carefully in the economic evaluation through the different approaches considered in Sect. 2.3.1. The intermodal secondary effects can be positive or negative depending on the sign of the distortion and the cross elasticity, but in the case of optimal pricing, like road congestion pricing or optimally designed airport congestion charges, there are no additional benefits (or costs) in these markets.

Changes in proximity derived from transport investments can cause increases in productivity through different mechanisms linked to the economies of agglomeration or changes in land use (see Graham & Gibbons, 2019; Laird et al., 2014; Venables, 2007). These are the WEBs, and the risk, in this case, is to confuse relocation with growth. Relocation occurs when some benefits of the project come from deviation of the economic activity somewhere else, without any change in productivity, while growth occurs when the project adds value to the economy.

The three sources of WEBs (imperfect competition, tax revenues arising from labour market impacts and agglomeration economies) have not received the same attention in the economic evaluation of projects. The focus has been directed to agglomeration economies because they are considered the main source of WEBs and also because their econometric estimation is easier (see Graham & Gibbons, 2019).

From this line of reasoning, it is clear that there may be consequences of transport investment that relate specifically to agglomeration. The rationale is that if there are increasing returns to spatial concentration, and if transport improvements partially explain the level of concentration or density experienced by firms and workers, then investment in transport may induce productivity gains thanks to the positive externalities of agglomeration.

Conventional CBA may underestimate the benefits of large infrastructure projects if economies of agglomeration are significant. This calls for its inclusion together with the direct user benefits. At the same time, the recent popularity of WEBs, as well as the use of impact studies, may also be revealing the interest of promoters to get the approval of projects with modest direct benefits. There are some general principles to avoid a mechanical and may be misleading use of WEBs (Venables, 2019): (i) Narrative: there should be a clear narrative of the main problem that policy is intended to address and the key market failure(s) that motivate the policy; (ii) transparency: the mechanisms underpinning both the quantity changes and their social value should be clear and explained in a manner that enables the key magnitudes to be understood from straightforward back-of-the-envelope calculation; (iii) sensitivity: there should be an analysis of the dependence of the quantity effects and their valuation on key assumptions about the economic environment. Scenarios outlining the quantitative importance of failure of these assumptions should be outlined; (iv) complementary policies: there should be a thorough consideration of complementary measures that are needed for a successful implementation of a project; (v) alternatives: any project should make a strong case that it provides the most cost-effective way to solve the main problem described in the narrative.

A practical approach for small projects is to concentrate the effort in the identification and measurement of direct effects. Although this approach faces the risk of ignoring WEBs in the case where they exist and are significant, there is consensus on the fact that this is a trade-off between the risk of ignoring a real effect and the risk of double counting and unnecessary delays in project evaluation. For large projects or for the evaluation of investment programs, it may be justified to undertake more complex analyses. Nevertheless, there is some confusion on whether some studies are measuring additional economic effects, ignored in standard CBA, or whether they are only providing the final impact of the direct effects already measured.

From the available empirical evidence and the evaluation of the experts on whether conventional CBA is sufficient to estimate the social profitability of a project, the general recommendation is to be extremely cautious since, although economists are advancing with the knowledge and measurement of WEBs, they are still far from turning the results into practical rules for their inclusion in CBA. It does not seem reasonable to transfer the results from other studies, using conversion factors or similar procedures, considering the variability of the value of WEBs, and even the sign when there are some negative effects like congestion and similar externalities. Moreover, when investment induces agglomeration, it may also induce additional negative externalities not fully captured in the analysis, as the negative political and social consequences of territorial inequalities (Rodriguez-Pose, 2018).

2.4 The Relationship between Pricing and Investment

Prices affect demand and, thus, the social surplus of the investment project. Therefore, before evaluating a transport infrastructure, we need to know the associated charging scheme. This is the main argument of Dupuit (1844) when analysing the social welfare of a toll-free bridge. If there is no charge for the use of the bridge and there only fixed costs, the bridge should be constructed if (the maximum) consumers' surplus is higher than the construction cost. If the regulator charges a price for the use of the bridge, consumers' surplus decreases, and this change in the charging scheme reduces the social surplus, leading to a situation in which it may well happen that the socially optimal decision is now not to build the bridge.

Making decisions about accepting or rejecting projects always requires taking into account how external interventions in transport markets affect the successive equilibria that occur in these and other related markets. This is hardly compatible with the assumption that prices are completely exogenous, and therefore, we have to be aware that the social profitability of any given project may differ under different pricing policies. Prices affect demand and, thus, the size (or even the sign) of agents' surpluses and whether the revenues will be enough to cover costs. Therefore, when deciding whether it is socially desirable to invest in a project or not, its economic evaluation should consider the future charging scheme (including the cases where no prices are charged at all) and whether the project requires the use of public funds.

Let us consider as an example a transport project consisting in building a new transport infrastructure under the assumption that there are no other transport alternatives to address the same (mobility) problem.[20] We will first assume that there is no economic cost of public funds, which is defined as the loss incurred by society when raising revenues to finance government spending through distortionary taxation. Secondly, we will discuss the role of the pricing scheme and the economic cost of public funds in the investment decision. In all cases, we will assume that there is enough capacity to satisfy the demand, i.e., there are not scarcity or congestion.

2.4.1 Investment Decisions When There is no Economic Cost of Public Funds

Consider the evaluation of a new transport infrastructure that requires an initial investment denoted by I. In order to illustrate the importance of the pricing scheme on this investment decision, throughout this section, we will define as $\Delta \overline{W}$ the change in social welfare due to the project as the sum of consumers' and producers' surpluses[21] net of the investment cost.

The new infrastructure should be built if the net increase in social welfare, $\Delta \overline{W}$, is higher than the investment cost, I:

$$\Delta \overline{W} = \Delta CS + \Delta PS > I. \tag{2.62}$$

Let us start assuming that there are no operating costs, there are no other transport alternatives, and there is no economic cost of public funds. Figure 2.2 represents the inverse demand function for the use of such a new transport infrastructure.

---

[20] See Chap. 4 for the analysis of pricing and investment when there are other transport alternatives.

[21] Notice that producers' surplus refers to both capital owners' surplus and land owners' surplus, and all inputs are paid at their marginal opportunity cost.

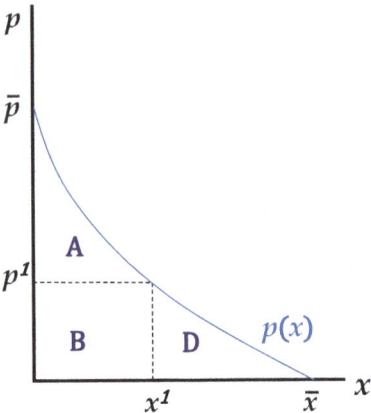

Fig. 2.2 Pricing and social welfare when there are no operating costs

The increase in social welfare due to this transport project depends on the pricing scheme. If there is free access to the transport infrastructure ($p = 0$), the net change in social welfare is equal to the increase in consumers' surplus due to the project (given by $A + B + D$), since producers' surplus is equal to zero. Although, in this case, free access is the optimal pricing scheme, the government may be budget-constrained and a positive price for the use of the new infrastructure may be needed.

If this price is equal to p^1, the net change in social welfare is given by the increase in consumers' surplus (A) and the increase in producers' surplus (B). Therefore, the social profitability of the project critically depends on the pricing scheme, since:

$$\Delta \overline{W}_{p=0} = A + B + D,$$
$$\Delta \overline{W}_{p=p^1} = A + B. \tag{2.63}$$

How does the charging scheme affect the investment decision? If $A + B > I$, the optimal decision is to accept the project, regardless on whether the government plans to charge p^1 or allow free access. On the contrary, if $A + B + D < I$, the optimal decision is not to invest independently on the charging scheme. Interestingly, it might be the case that $A + B + D > I > A + B$, i.e., under a free access scheme, the infrastructure should be constructed, but if the government is planning to charge p^1, the socially optimal decision is not to invest in this transport infrastructure. Therefore, the investment decision may strongly depend on the pricing scheme to be used once the infrastructure has been constructed.

Now consider that marginal operating costs are constant and equal to $c > 0$, as in Fig. 2.3. If there is free access, the net change in social welfare is equal to the increase in consumers' surplus, given by areas $A + B + D + E + F + G$ plus the decrease in producers' surplus, given by $E + F + G + H$. Thus, the net change in social welfare is $A + B + D - H$. Conversely, if the government applies an optimal marginal cost pricing scheme ($p^0 = c$), the net change in social welfare is equal only

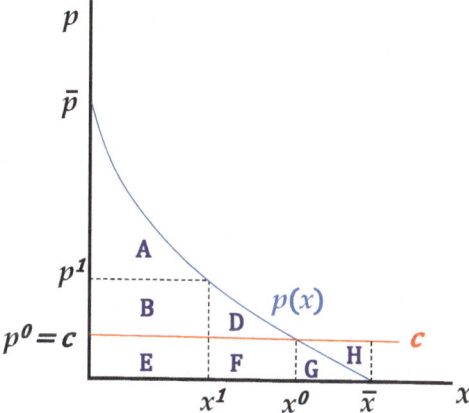

Fig. 2.3 Pricing and social welfare with constant marginal operating costs

to the increase in consumers' surplus $(A + B + D)$, since, in this case, producers' surplus is equal to zero. Finally, if the price is set above marginal costs, $p^1 > p^0$, the net change in social welfare due to the transport project is given by the increase in consumers' surplus (A) and the increase in producers' surplus (B). These results can be summarized as:

$$\Delta\overline{W}_{p=0} = A + B + D - H,$$
$$\Delta\overline{W}_{p=c} = A + B + D,$$
$$\Delta\overline{W}_{p=p^1} = A + B, \tag{2.64}$$

showing that, once again, the optimal investment decision according to expression (2.62) may strongly depend on the pricing scheme. Suppose, for example, that the values in the figure are such that $A + B + D > I$, but $A + B + D - H < I$ and $A + B < I$. In this case, it is socially optimal to invest in the construction of the new infrastructure if the government is planning to charge a price equal to marginal operating costs, but the project should be rejected if free access or prices higher than marginal operating costs are being considered as pricing policies.

2.4.2 Investment Decisions When There is an Economic Cost of Public Funds

Let us now consider that public funds are obtained through distortionary taxation and, thus, there is a shadow price of public funds denoted by $\lambda_g > 1$. This means that when the government uses I monetary units of public funds to build the new transport infrastructure, this investment really costs $\lambda_g I$ to the society in terms of the distortions created in the economy when raising taxes to finance such an investment. However, if the government charges a positive price for the use of the new infrastructure and collects net revenues equal to $(p - c)x$, this amount can be used to reduce existing distortionary taxation, and thus, the real profit for the society is $\lambda_g(p - c)x$.

Therefore, defining the net increase in social welfare due to the transport project as the sum of consumers' surplus and producers' surplus (net of the investment cost), the 'acceptability condition' for the new infrastructure becomes:

$$\Delta \overline{W} = \Delta\text{CS} + \lambda_g \Delta\text{PS} > \lambda_g I. \tag{2.65}$$

Again, let us start the discussion assuming that there are no operating costs, as in Fig. 2.2. If there is free access to the transport infrastructure, the only net change in social welfare due to the project is equal to the increase in consumers' surplus $(A + B + D)$, since producers' surplus is equal to zero. On the contrary, if the government charges a price p^1 for the use of the infrastructure, the net change in social welfare is given by the increase in consumers' surplus (A) plus the increase in producers' surplus (B multiplied by the shadow price of public funds). Once more, the social profitability of the project depends on the pricing scheme:

$$\Delta \overline{W}_{p=0} = A + B + D,$$
$$\Delta \overline{W}_{p=p^1} = A + \lambda_g B. \tag{2.66}$$

How does the shadow price of public funds affect the investment decision? If λ_g is high enough, it might be that $A + \lambda_g B > \lambda_g I > A + B + D$. In this case, if there is free access to the transport infrastructure, the socially optimal decision is to reject the project, but if the government is planning to charge a price p^1, the socially optimal decision is to build the infrastructure.

Suppose now that marginal operating costs are constant and equal to c as in Fig. 2.3. Under a free access pricing scheme, the net change in social welfare is again equal to the increase in consumers' surplus $(A + B + D + E + F + G)$ minus the decrease in producers' surplus, given by $\lambda_g(E + F + G + H)$. If the price is $p^0 = c$, the net change in social welfare is just equal to the increase in consumers' surplus $(A + B + D)$, because producers' surplus is equal to zero. Finally, if the government charges p^1, the net change in social welfare is given by the increase in consumers' surplus (A) and the increase in producers' surplus $(\lambda_g B)$. In sum, we have that:

$$\Delta \overline{W}_{p=0} = A + B + D + (1 - \lambda_g)(E + F + G) - \lambda_g H,$$
$$\Delta \overline{W}_{p=c} = A + B + D,$$
$$\Delta \overline{W}_{p=p^1} = A + \lambda_g B. \tag{2.67}$$

If the shadow price of public funds is large enough, we could have, for example, that $A + \lambda_g B > \lambda_g I > A + B + D$. In this case, it would be socially optimal to invest in the construction of the new infrastructure if the government is planning to charge prices above marginal operating costs, but the transport infrastructure should not be built if the government is planning to allow free access or charge a price equal to marginal operating costs. So, once again, we have shown that the investment decision may strongly depend on the pricing scheme.

2.5   Conclusions

The economic evaluation of transport projects is well developed, and there is a rich sample of applications by transport modes and policies. This chapter aims to stress the importance of following an analytical approach in the evaluation of transport projects, instead of the mechanical application of the common rules of thumb applied in national or supranational CBA guidelines.

A transport project can be contemplated as an external intervention moving the economy from one equilibrium to another. The project changes generalized prices and quantities in the transport markets as well as in other secondary markets. The economic appraisal of projects tries to estimate the net welfare effect of the intervention instead of impacts on gross domestic product or employment, and for these reasons, effects in secondary markets without distortions and any other effects (e.g., income multiplier) common to the counterfactual can be ignored in the CBA of the project.

The purpose of the CBA is to evaluate whether the expected social benefits exceed the opportunity cost of the resources diverted from other uses to obtain those benefits. This is the challenge of the economic evaluation of projects and policies, whose main objective is to assess changes in the well-being of individuals directly or indirectly affected by their implementation.

There are two key approaches for the calculation of the economic profitability of a project. One is based on the aggregation of the changes of the economic surpluses of the different groups in the society. Alternatively, the practitioner may ignore transfers between groups and concentrate on the change in WTP and the use of resources with the project compared with the counterfactual. Both approaches lead to the same result, but once the practitioner chooses one of them, the method must be strictly followed.

It is quite frequent that, due to lack of data and the difficulty to ex-ante identify the final beneficiaries, the second approach is followed. In this case, there are two alternative options, especially relevant when, as usual, other modes or activities are subject to distortions, such as taxes or market power. These options are what we call the single graph analysis (unimodal), or the corridor analysis (multimodal). In the first one, the analyst can concentrate the effort on the primary market, adding an estimation of the effects on taxes or market power in the secondary markets. In the second one, the multimodal or the corridor analysis, these effects can be overlooked and only changes in resources within the transport corridor are accounted for. Therefore, when generated demand is not significant or the price is equal to the social marginal cost in other transport modes or activities, the single graph is a good option. However, when there exist taxes, externalities or market power in other transport modes or activities, the corridor analysis is highly recommended.

We derive the rules for the measurement of the effects of transport projects from a simple mode including time in the generalized prices of goods and services, and the corresponding budget constraint given the time endowment of the individual. In a nutshell, before applying any set of practical rules for the economic evaluation of any project, the practitioner should know the analytical model behind those rules, and whether the application of those rules is consistent with such a model, avoiding the combined use of different rules from different approaches.

There are other potential effects of transport projects beyond the direct benefits. This is the case of the WEBs derived from the response of the private sector to changes in proximity. Depending on the context, these effects could be significant in affecting productivity and land development. There are many *small* projects where these effects can be safely ignored. In *large* projects where the effects are expected to be significant, like in transformational urban infrastructure projects clearly associated with agglomeration economies, they should be considered both in their positive aspects (increase in productivity) as well as in their negative externalities (productivity losses in the periphery and congestion in the centre).

Finally, we have to take into account that prices affect demand and, thus, the social welfare of transport projects. Therefore, when deciding whether it is socially desirable to invest in a transport project or not, its economic evaluation should consider the future charging scheme (including the cases where no prices are charged at all) and whether the project requires the use of public funds.

Chapter 3
Cost–Benefit Analysis of Transport Investments: An Application to High-Speed Rail Investments

3.1 Introduction

There are many social benefits of high-speed rail (HSR). Although they are not always easy to identify and measure, they typically include time savings for passengers, who can travel faster, and with better service quality, between destinations; fewer cars on the road, which can reduce congestion, accidents and air pollution; less intensive use of air transport; and reduced dependence on fuel and oil. Their promoters also argue that building HSR lines boosts economic growth, increasing productivity in cities and urban areas by bringing distant population closer together. Nevertheless, these effects cannot be generalized (see Albalate & Bel, 2017; de Rus, 2023; Preston, 2017).

However, most HSR projects require significant investment in infrastructure, rolling stock and maintenance. Operating costs can also be high due to the need for skilled personnel and energy consumption, and they can vary considerably from one project to another. The existence of pre-existing infrastructure that can be upgraded, or the costs associated with site acquisition and preparation, can also significantly alter the social and private costs of these projects (including the environmental costs of the required public works).

All these reasons justify the need to carry out thorough feasibility and cost–benefit analysis (CBA) studies before embracing HSR projects. Although their economic assessment is based on the same economic principles that apply to any other mode of transport, the general rules should also consider the technical and economic specificities of a sector characterized not only by its ability to provide fast and reliable services to large numbers of passengers and freight but also by its high (and sunk) construction, operation and maintenance costs (de Rus, 2012; de Rus & Nash, 2009; Nash, 2014; Preston, 2013).

Rail transport is also characterized by large economies of scale and density, which have traditionally been used in many countries to justify market structures based on vertically integrated public monopolies with limited opportunities for competition. In recent decades, in order to reverse the gradual decline of rail transport in relation to other competing modes, this structure has changed in Europe. The European Union (EU) has embarked on a restructuring process based on the separation of infrastructure management from service provision and the progressive liberalization of the sector to encourage competition either 'on the tracks' or 'for the tracks'. The increasing number of public and private partners currently involved in the rail sector makes it even more important to properly quantify and analyse the distribution of benefits and costs between them when undertaking any new project, and indeed it becomes an essential part of the overall decision-making process (Beria et al., 2012; Gomez-Ibañez & de Rus, 2006).

Although in most EU Member States government remains responsible for many rail investments and policies, the economic evaluation of rail projects has to take into account not only their impact on the public sector budget but also any significant economic impact in the society: the infrastructure manager, the service operators (public and private), the users of rail, and other strongly interrelated markets (basically the other modes of transport). Although some smaller projects may still be decided at the level of the infrastructure manager or operators, larger projects (e.g., opening a new railway line, upgrading or closing existing services, building new infrastructure, etc.) are increasingly affected by these complex relationships. Decisions on these projects are becoming more difficult and require a more comprehensive assessment based on sound and well-established economic criteria.

In this chapter, we focus on the main features of the CBA of HSR projects, although, due to the similarities between conventional train and HSR in their key elements and demand forecasting, we start in Sects. 3.2 and 3.3 examining general rail projects. The peculiar characteristics of HSR have won the support of users (quality and speed), governments (territorial equity, less contamination and congestion of airports and roads), rail authorities (a path of renewal in a context of declining market share), and firms (that produce rail equipment). But the problem is still economic profitability. The infrastructure costs are high, sunk, and associated with strong indivisibilities. The size of the infrastructure is essentially the same for any given line, regardless of the level of existing demand. Its use is associated with very pronounced decreasing average costs. For these reasons, per capita income, population density, and the state of the alternative mode of transport largely determine the financial and social viability of the investment.

The question is whether the society is willing to pay the social cost of the investment. This is the social value of the economic evaluation of projects. Market shares, the political rhetoric about its role in territorial integration, and its alleged positive impact on the environment and on regional development are no sufficient conditions for the social viability of the investment.

This is the main focus of this chapter, following the analytical evaluation model of Chap. 2, where a transport project is interpreted as a perturbation in the economy affecting the social welfare of different individuals at different moments in time, as compared with the counterfactual. Using that approach, Sect. 3.2 starts by discussing how rail transport projects should be defined from the point of view of their ex-ante assessment, particularly focusing on the different roles played by the different agents in the sector. The rest of the section focuses on passenger rail undertakings (building a new line, expanding an existing network, etc. Section 3.3 analyses the importance of demand forecasts in rail projects. Section 3.4 examines in more detail some of the most relevant technical features of these projects, particularly for interurban HSR lines. Section 3.5 discusses the application of the CBA methodology of these rail projects, providing a hypothetical case to illustrate the assessment process. It also provides a comparison of the approaches to project appraisal shown in Chap. 2 as well as the equivalence between them. Finally, Sect. 3.6 analyses additional economic benefits of HSR.

3.2 Defining a Rail Project Assessment

UIC (2022) estimated that, by the end of 2021, there were almost 5000 high-speed trains in operation around the world every day, carrying more than 2 billion passenger-trips a year, not only in countries such as China, Japan, France, Spain, Italy, or Germany, but also in Turkey, Morocco, the USA, Iran, Russia, India, and many others. The total length of the built network was almost 59,000 km, with a further 20,000 km under construction and almost 50,000 km under study.

As with any other transport project, the economic appraisal of these rail investments should start with the appropriate identification of the project within a wider investment program at regional, national, or European level. This first step defines the scope of the analysis and clarifies the stakeholders in the evaluation. From a technical point of view, the project should then be defined in terms of all the (technical) elements needed to make it operational (e.g., main and secondary lines, stations, depots and other ancillary infrastructure, energy and communication facilities, rolling stock, etc.). Moreover, from an economic point of view, the project definition should avoid unrelated elements or those not necessary to make the project operational (e.g., buildings not related to train operation, roads not required by/for the railway project, etc.).

Most (large) passenger rail projects require prior planning to clarify what the final objective is and how it can be achieved. For example, some projects may be designed to solve very specific problems (e.g., bottlenecks or lack of capacity); others may be designed to improve current transport conditions (e.g., slow connections or poor quality of service); and in many cases, they may respond to other societal needs (e.g., improved accessibility, reduced environmental impact). Since undertaking a project involves a simultaneous decision not to undertake any of the other feasible options, a reasonable range of alternatives should always be considered in order to assess the economic viability of a project.

One of the options to be considered is the 'do-minimum', which implies carrying out as little investment and maintenance as possible to keep current transport markets functioning without undue degradation of service. In the case of railways, this can be interpreted as following the standard pattern of renewing and maintaining the existing infrastructure and rolling stock (which, of course, would result in significantly different traffic levels than those foreseen in the project).

On the contrary, the 'do-nothing' alternative is often incompatible with normal operation on the existing network and is therefore usually not a valid reference. In many cases, particularly in the evaluation of high-speed lines, the 'do-minimum' is not the investment needed to provide the capacity required by expected normal traffic growth (referred to as 'avoided investment'), but an 'avoided' major alternative (such as track doubling). In cases where saturation of the conventional rail network requires capacity expansion, the construction of a new HSR line should also be considered as an alternative to improving and expanding the conventional network, with the added benefit of freeing up capacity. Obviously, this additional capacity is valuable if demand exceeds existing capacity. In these circumstances, additional capacity can absorb traffic growth between cities served by HSR and also free up capacity on existing lines to serve other traffic (such as suburban or freight demand).

Thus, there may be 'do-something' alternatives that can be defined in a variety of ways, depending on the project size and scope. Sometimes, an alternative is simply an extension or improvement of another (e.g., an additional link). In this case, if the basic alternative is acceptable, it is the extra investment what must be appraised. Comparisons become more complex when several interlinked projects are evaluated. If network effects are relevant (e.g., in the case of suburban railways), the implementation of related projects and their timing could have important effects on the profitability of the whole investment. One possible way to handle such cases is to carry out appraisals of the whole investment and of each of its individual components, to reach both an optimal project selection and their scheduling period. However, this is quite difficult in practice, and an individual appraisal of each project is performed.

In fact, the traditional view when defining (particularly, urban and suburban) rail projects has been to consider investments leading to a continuous improvement process rather than options representing a major change in the network. This is due to the integrated character of the rail system that often prevents the spread of advantages (notably those derived from innovation) throughout the whole network. Speed restrictions or old electrification and signalling systems can, for instance, make inefficient the deployment of modern rolling stock on upgraded sections. However, the increasing development of high-speed networks (which require dedicated infrastructure) is changing this approach, and some experts argue for slightly different assessment approaches to different rail projects (UIC, 2018). In all cases, the definition of alternatives for a rail project should always consider the implications for the whole transport system.

Finally, the investment decision of building a HSR network should be subject to a proper consideration of the alternative outside the rail market. This is specially so in the case of countries and corridors with a well-established airport network and competitive air services, low demand, and tight budget constraints. A social profitable new HSR line requires a volume of users, and willingness to pay (WTP), much higher than the current ones in many of the existing or planned corridors in Europe. Although also subject to indivisibilities, airports are more adjustable than railways to population density. This is not the case with railways, where indivisibilities imply a very high average cost with low demand (de Rus & Socorro, 2017).

3.3 Demand Forecasting and Rail Project Evaluation

Once the project has been defined, the next step in the evaluation process is to carry out a detailed demand analysis. This involves assessing the existing demand for the different types of users (typically using data provided by current service providers or statistical offices) and the future demand (based on reliable models that take into account reasonable economic forecasts, alternative sources of supply, demand elasticity to relevant prices and incomes, etc.). Both 'with the project' and 'without the project' passenger figures are essential for the formulation of projections, although the demand analysis should always provide forecasts adapted to the technical and economic characteristics of each project, including the impact of pricing policies and the foreseeable response of all relevant stakeholders to changes in generalized prices. Depending on the available data and resources, different techniques (e.g., regression models, logit models, trend extrapolation, qualitative methods, etc.) can be used to forecast demand.

In general, as transport projects affect the modal distribution of demand, it is important to clearly identify the sources of deviated demand (from competing modes to rail) and generated demand by the rail project. It is important to distinguish between travel purposes (business, commuting, leisure, etc.), as users value time differently, and between routes or specific sections. Particular attention should be paid to whether the project is part of a network (as in the case of many suburban rail projects), as its demand is strongly influenced by issues of complementarity and accessibility, which should be explicitly considered in the appraisal.

In practice, although there are different procedures for carrying out demand analysis in rail projects, they all typically include at least three elements. For example, when considering the construction of a new passenger rail line (e.g., a new high-speed corridor), it is necessary to:

1. Build a database, usually in the form of a detailed origin-destination (OD) matrix, that identifies the most relevant OD relationships between the cities/stations affected by the project. This matrix should include all the available data (existing demand, number of services, frequencies, average speed, monetary prices, etc.) for all the current transport alternatives (conventional train, air transport, bus, and car), as well the most relevant parameters (e.g., unit costs estimates) required for an ex-ante CBA.

2. Estimate the modal distribution, in particular the generation and distribution of new and existing trips according to a generalized price demand model that determines what happens with and without the project. The resulting modal split must be consistent with all the economic and technical parameters of the model (e.g., prices and capacity restrictions).

3. Forecast the evolution of the demand over the evaluation period, which should be defined in accordance with the economic life of the assets involved in the project. For investments in rail infrastructure, a reasonable horizon lies between 30 and

50 years, although the shorter the period, the higher the residual value to consider at the end of the evaluation. Demand forecasting should be based on estimated gross domestic product projections and the corresponding income elasticity.

The use of generalized prices is the central element in demand prediction models in the CBA of transport projects. By reducing travel time, rail becomes more attractive compared to its road and air competitors, especially for medium distances. Average speed is thus a key determinant of users' modal choice. For example, rail market shares (calculated with respect to the combined air–rail market) in corridors where travel time is below 2.5 h are above 80%, while for routes where it exceeds 3.5 h they often fall below 50% (Campos & Gagnepain, 2009; UIC, 2018). Similarly, increases in frequencies lead to lower waiting times.

In the demand analysis of urban and suburban rail projects, other factors are also considered prior to proceed with the assessment process. Of particular interest in several European countries is the fact that most suburban rail services in metropolitan areas are subject to public service obligations, and existing regulations set up a series of prerequisites that must be fulfilled in order to authorize new projects.

Finally, there are several additional factors that are relevant to the demand analysis of rail projects and should be included in the project appraisal process when the information is available. The sensitivity of demand estimates to, for example, demographic and socio-economic changes (in the characteristics of users, their travel preferences and their WTP), the industrial and logistical structure of the area concerned (location of economic and social activities), the strategies of competing modes of transport, government policies (including subsidies and taxes), and, of course, the pace of technological change, with particular implications for the rail industry in the current context of the development of some alternatives (levitation trains, hyperloop, etc.) that could render current rail technology obsolete in just a few decades.

3.4 The Costs of Building and Operating High-Speed Rail Projects

HSR is a ground-based, guided, and low-gradient transport system comprising at least three different technical elements: upgraded infrastructure or new lines designed for a maximum speed of 250 km/h or more; dedicated rolling stock with *ad hoc* designed trainsets; and new operational rules, communication, and maintenance systems that allow the provision of high-quality passenger transport services. Although this commercial speed is the main reference for HSR, on medium distance lines without competition from air transport, lower average values (but above 200 km/h) are acceptable in Europe if the services include specific trains, no track-side signalling, long-distance control centres, and geographical or temporal separation of freight and passenger demand.

The Construction Costs of a High-Speed Line

According to the relationship between the infrastructure to be built with the pre-existing rail infrastructure, there are at least five different types of HSR projects: large corridors isolated from other lines, network integrated corridors, smaller extensions or new sections of existing corridors, large singular projects, and smaller projects complementing the conventional network, including high-speed lines that connect airports with nearby cities or the improvements in conventional infrastructure to accommodate higher-speed services.

In general, building an HSR infrastructure in any of these cases requires a specific design in order to remove all technical issues that might reduce trains commercial speed, including roadway-level crossings, sharp curves, excessive gradients, and other orographic limitations. For this reason, it is difficult to compare the construction costs of different HSR projects. In many projects, land and planning costs, and the cost of main stations, are excluded from the infrastructure construction costs and, around the world, the average cost of an HSR line per kilometre ranges, approximately, from €15 to €60 million in 2020 prices (see Campos & de Rus, 2009). Equivalent values can be found in Grob & Craven (2017), and UIC (2018).

Most high-speed lines are built within five to six years of taking possession of the necessary land unless tunnels and viaducts are numerous or long (UIC, 2018). The distribution of construction investment over this period is not uniform and depends on technical, economic, and even political factors. Although the administrative procedures may vary slightly from one country to another, once the decision to build a high-speed line has been taken, they usually include public enquiries, pre-project environmental studies of the areas affected, including a list of possible mitigation measures if necessary, and the design of the institutional and financial scheme to determine who will carry out the project and how it will be financed within the appropriate regional, national, or European framework.

Estimating the Costs of Operating and Maintaining a High-Speed Line

Once the infrastructure has been built, the operation of HSR involves two types of costs: those related to the operation and maintenance of the infrastructure itself, and those related to the provision of transport services. The degree of vertical integration existing between the infrastructure manager and the firm(s) that provides rail transport services determines which one is responsible for each of these costs and may vary across countries, making comparisons always difficult.

The first cost category includes labour, energy, and material costs associated with traffic management and the operation and maintenance of guideways, terminals, stations, signalling, and other ancillary systems. Some of these costs are fixed and depend on routine operations carried out in accordance with technical and safety standards. In other cases, such as track maintenance, the cost is determined by the intensity of traffic, which reflects the wear associated with the vertical forces carried by the track and its deformation due to excessive settlement of embankments or damage to structural elements. Therefore, these costs are conditioned by the demand that supports the line (number of trains per year). The maintenance cost functions follow a similar pattern, as the wear caused by trains also depends on the traffic volume. Signalling and safety systems costs can also be approximated in a similar way, while for stations, their costs are proportional to their size and intensity of use. In general, most of these infrastructure operation and maintenance activities require specialized staff, and labour costs usually account for about 50% of the total costs (Campos & de Rus, 2009). According to UIC (2018), the average maintenance costs for one kilometre of a new high-speed line can be estimated at €90,000 per year.

The costs associated with the provision of HSR services can be divided into three main categories: acquisition of rolling stock, operation and maintenance of rolling stock (mainly labour costs) and overheads. In order to estimate the provision of train services required to meet the expected demand, the assessment process should be based on an operational plan that defines how the new line will be operated. In order to size the train fleet, it is necessary to know not only the annual demand of the line, but also whether the technical characteristics of the trains allow them to operate within the specific design parameters of the line. The operating plan must, therefore, determine two aspects: the number of trains required each year (including replacements) and their type (in terms of capacity, maximum speed, architecture, power, traction, gauge, axle weight, signalling, etc.). In Europe, acquisition costs of a 350 trainset could reach €30–35 million, according to UIC (2018). Technical factors such as composition, mass, weight, power, traction, tilting characteristics, or internal configuration may affect this price.[1]

Deciding the number of trains on a particular corridor requires estimating the daily demand in each direction using annual and monthly estimates, once corrected by seasonality coefficients in order to avoid service disruptions and minimize idle capacity. Then, it is relatively simple to calculate the number of services per direction by dividing daily demand over the number of seats on the train multiplied by the target load factor set by the operator. From the daily services, the frequency and the number of trains for a basic service can be easily obtained from simple calculations, adding a number of extra trains (contingency factor) for replacing those in maintenance, rotations, repositioning, or breakdowns.

For rolling stock operating costs, unit cost estimates vary according to the line's operating plan. According to UIC (2018), the maintenance of a high-speed train (with an average annual use of 500,000 km) is estimated at €2 per train-km, i.e., €1 million per year. However, this amount could be significantly higher (at least tenfold) if we include operation costs (Campos & de Rus, 2009).

---

[1] An extensive review of the different types of rolling stock used worldwide and their specific characteristics can be found in UIC (2022).

3.5 Cost–Benefit Analysis of High-Speed Rail Projects

This section is devoted to the application of the methodology using a stylized case consisting of the construction and operation of a new HSR connecting two cities, with no intermediate stations, and the closure of the conventional railway services.

The net social benefit of the project is calculated with two different, but equivalent, approaches (see Chap. 2): adding the changes in the surpluses of the agents affected by the project, and adding the change in the WTP for the project and in the use of real resources, ignoring transfers. In the second approach, we consider two options: the single graph analysis (unimodal) or the corridor analysis (multimodal).

This project is a simplified case of a high-speed line evaluation. The value of the parameters used in this practical example is only for illustration purposes. The results obtained in this illustration critically depend on the chosen values. The appraisal framework can easily be used for actual projects.

3.5.1 Methodology to Compute the Social Welfare Effects of a High-Speed Rail Project

In order to evaluate the HSR project, we depart from the model developed in Chap. 2. We assume that the evaluation of the effects of the investment in the group of interrelated markets of the different transport modes is enough to obtain an approximation to the net welfare effect of the project, as explained in Chap. 2.

We consider a market with n modes of transport or activities, where the transport project reduces the generalized price of conventional train ($g_i^1 < g_i^0$) due to a reduction of in-vehicle time. The superscript 1 indicates the situation with the project and superscript 0 denotes without the project; and the generalized price includes the monetary price paid by users, p_i, plus the corresponding time cost $v_i t_i$, where v_i is the value of time of users initially travelling in conventional train and t_i the total travel time of this mode. Time cost is disaggregated in three components: in-vehicle, waiting, and access and egress time. We also denote by τ_i the value of an *ad valorem* tax applied in the rail market. Notice that, since there is an *ad valorem* tax, the price charged by producers (p_i^-) does not coincide with the price paid by users (p_i), where $p_i = p_i^- (1 + \tau_i)$.

In the case of deviated demand from mode or activity j,[2] g_j^0 denotes the generalized price for the user indifferent between mode or activity j and conventional train without the project, that includes the monetary price paid by users, p_j, plus the corresponding time cost, $v_j t_j^0$, where v_j is the value of time of users initially consuming mode or

---

[2] Recall from Chap. 2 that we may distinguish between existing demand (conventional train), deviated demand, i.e., changing from an alternative mode (plane, bus or car) and generated demand, i.e., coming from other activities.

activity j and t_j the total travel time of this alternative, that is disaggregated in the three aforementioned components. Recall that the indifferent user has the same generalized price in mode or alternative j and in conventional train, so $g_j^0 = g_i^{0d} = p_i^0 + v_j t_i^0$. Moreover, once the project is implemented, the generalized price in the rail market is reduced to $g_i^{1d} = p_i^1 + v_j t_i^1$ and, due to this reduction, some users that preferred mode or activity j without the project now prefer HSR.[3] Now, there is a new indifferent consumer, and his generalized price in the alternative is, $g_j^1 = p_j + v_j t_j^1$ and, similarly to the former indifferent user, in the final equilibrium, $g_j^1 = g_i^{1d}$ for the new one.

Moreover, there are *ad valorem* taxes in all the alternatives (τ_j), so the price charged by producers (p_j^-) does not coincide with the price paid by users (p_j), where $p_j = p_j^- (1 + \tau_j)$, with $j = 1, \ldots, n$ and $j \neq i$.

In order to obtain the change in social welfare of this transport project, we must compute the social net present value (NPV$_S$) of the project by adding the discounted changes in social welfare over the evaluation period ($t = 0, \ldots, T$) using the corresponding social discount rate (r_S):

$$\text{NPV}_S = \sum_{t=0}^{T} \frac{\Delta W_t^e + \sum_{\substack{j=1 \\ j \neq i}}^{n} \Delta W_{jt}^d}{(1 + r_S)^t}, \tag{3.1}$$

where $\Delta W_t^e + \sum_{\substack{j=1 \\ j \neq i}}^{n} \Delta W_{jt}^d = \Delta W_t$ denote the change in social welfare in year t, and superscripts e and d denote changes due to existing demand and deviated and generated demand from mode or activity j, respectively. For the sake of simplicity, from now on the subscript t is omitted, and we focus on the approaches to compute the change in social welfare every year.

First, let us start with the assessment of the welfare effects of the project using the methodological approach based on the change in agents' surpluses affected by the project.[4] The changes in users' surplus for existing demand (ΔCS^e) and for deviated and generated demand $\left(\sum_{\substack{j=1 \\ j \neq i}}^{n} \Delta CS_j^d \right)$ must be obtained using tax-inclusive prices, according to the following equations:

$$\Delta CS^e = \left(g_i^0 - g_i^1 \right) x_i^0, \tag{3.2}$$

$$\sum_{\substack{j=1 \\ j \neq i}}^{n} \Delta CS_j^d = \sum_{\substack{j=1 \\ j \neq i}}^{n} \frac{1}{2} \left(g_j^0 - g_i^{1d} \right) x_j^d. \tag{3.3}$$

---

[3] We assume that total travel time changes with the project for deviated demand from other modes or activities because consumers have different access and egress time.

[4] We assume no change in workers' surplus nor landowners' surplus.

where x_i^0 and x_j^d are the existing demand travelling in conventional train and the deviated demand from mode or alternative j, respectively. Thus, total demand with the project (x_i^1) is equal to $x_i^0 + \sum_{\substack{j=1 \\ j \neq i}}^{n} x_j^d$.

Changes in capital owners' surplus are assumed to be zero for all alternative transport modes and activities. In the case of the rail market, they are calculated as the difference between the new revenue for the HSR operator (OP) minus, on the one hand, the acquisition, operation, and maintenance costs of the rolling stock (C_i^{RS}) and, on the other hand, the access charges (which represent a cost for the operator). For the infrastructure manager (IM), his surplus is given by revenues from access charges, minus the infrastructure investment cost (I) and his operation and maintenance costs (C_i^I). Therefore:

$$\Delta OS_{OP}^e + \Delta OS_{IM}^e + \sum_{\substack{j=1 \\ j \neq i}}^{n} (\Delta OS_{OPj}^d + \Delta OS_{IMj}^d) = p_i^{1-} x_i^0 + \sum_{\substack{j=1 \\ j \neq i}}^{n} p_i^{1-} x_j^d - C_i^{RS} - I - C_i^I. \quad (3.4)$$

Tax revenues from consumers are allocated to the change in taxpayers' surplus. The taxes paid by infrastructure manager because of the investment cost and his operation and maintenance costs (T^I), the taxes paid by HSR operator because of the acquisition and operation and management of rolling stock (T^{RS}), and taxes lost from other modes or activities, are also allocated to the change in taxpayers' surplus:

$$\Delta GS^e + \sum_{\substack{j=1 \\ j \neq i}}^{n} \Delta GS_j^d = \tau_i p_i^{1-} x_i^0 + \sum_{\substack{j=1 \\ j \neq i}}^{n} \tau_i p_i^{1-} x_j^d + T^I + T^{RS} - \tau_i p_i^{0-} x_i^0 - \sum_{\substack{j=1 \\ j \neq i}}^{n} \tau_j p_j^- x_j^d. \quad (3.5)$$

For simplicity, the external cost savings have been calculated under the assumption that the accident costs $(\mathcal{E})$ are not internalized and therefore allocated in the change in the surplus of the rest of society, i.e.:

$$\Delta ES^e + \sum_{\substack{j=1 \\ j \neq i}}^{n} \Delta ES_j^d = \mathcal{E}. \quad (3.6)$$

Thus, the change in social welfare is equal to:

$$\Delta W = \Delta CS^e + \sum_{\substack{j=1 \\ j \neq i}}^{n} \Delta CS_j^d + \Delta OS_{OP}^e + \Delta OS_{IM}^e + \sum_{\substack{j=1 \\ j \neq i}}^{n} (\Delta OS_{OPj}^d + \Delta OS_{IMj}^d)$$

$$+ \Delta GS^e + \sum_{\substack{j=1 \\ j \neq i}}^{n} \Delta GS_j^d + \Delta ES^e + \sum_{\substack{j=1 \\ j \neq i}}^{n} \Delta ES_j^d. \quad (3.7)$$

In the appraisal based on the changes in WTP and resources, we have two options: the single graph analysis (unimodal) or the corridor analysis (multimodal). As described in Chap. 2, the approach based in the changes in resources and in the WTP using the single graph analysis allows us to obtain the following effects:

1. Change in resources equals to the value of the time saved and the savings in operating cost associated with conventional train.

$$v_i\left(t_i^0 - t_i^1\right)x_i^0 + c_i^{0-}x_i^0, \tag{3.8}$$

where c_i^{0-} is the marginal operating cost of the conventional train, net of taxes.

2. Change in WTP for deviated demand (from air transport, bus, and car) and generated demand, minus the value of the time spent on the new trips.

$$\frac{1}{2}\sum_{\substack{j=1 \\ j\neq i}}^{n}\left(g_j^0 + g_i^{1d}\right)x_j^d - \sum_{\substack{j=1 \\ j\neq i}}^{n}v_jt_i^1x_j^d. \tag{3.9}$$

3. Moreover, there is a change in resources (the costs of the HSR, net of taxes): the acquisition, operation and maintenance costs of the rolling stock (C_i^{RS-}) and the infrastructure investment cost (I^-) and his operation and maintenance costs (C_i^{I-}), where $C_i^{RS} = C_i^{RS-} + T^{RS}$ and $I + C_i^I = I^- + C_i^{I-} + T^I$.

In sum, adding the changes in WTP and resources following the unimodal or single graph analysis, we obtain:

$$\Delta\text{Resources}^e + \sum_{\substack{j=1 \\ j\neq i}}^{n}\left(\Delta\text{WTP}_j^d - \Delta\text{Resources}_j^d\right) = v_i\left(t_i^0 - t_i^1\right)x_i^0 + c_i^{0-}x_i^0$$

$$+ \frac{1}{2}\sum_{\substack{j=1 \\ j\neq i}}^{n}\left(g_j^0 + g_i^{1d}\right)x_j^d - \sum_{\substack{j=1 \\ j\neq i}}^{n}v_jt_i^1x_j^d - C_i^{RS-} - I^- - C_i^{I-}. \tag{3.10}$$

4. Finally, we consider the effects due to distortions in other modes or activities. In our case, we have to take into account the taxes in other modes and activities and accidents cost savings:

$$\sum_{\substack{j=1 \\ j\neq i}}^{n}\left(-\tau_j p_j^- x_j^d\right) + \mathcal{E}. \tag{3.11}$$

Adding all the effects, the change in social welfare is given by the following expression:

$$\Delta W = \Delta \text{Resources}^e + \sum_{\substack{j=1 \\ j \neq i}}^{n} (\Delta \text{WTP}_j^d - \Delta \text{Resources}_j^d) + \sum_{\substack{j=1 \\ j \neq i}}^{n} \left(-\tau_j p_j^- x_j^d\right) + \mathcal{E}.$$

$$(3.12)$$

Finally, the approach based on the changes in resources and in WTP using the corridor analysis allows us to obtain the following benefits:

1. Value of time savings and cost saving for existing demand:

$$v_i\left(t_i^0 - t_i^1\right)x_i^0 + c_i^{0-}x_i^0.$$

$$(3.13)$$

2. Value of time savings and cost savings for deviated demand from other modes or activities. On the one hand, recall that the time saved by each consumer shifting from alternative j to HSR is not the same for everyone. Time savings for the indifferent consumer without the project are the highest while time savings for the new indifferent consumer with the project are the lowest (see Chap. 2 for a deeper explanation and the equivalences of the different expressions of time savings for deviated demand from alternative j). They could be computed as:

$$\sum_{\substack{j=1 \\ j \neq i}}^{n} \frac{1}{2}[((g_j^0 - p_j) - (g_i^{1d} - p_i^1)) + ((g_i^{1d} - p_j) - (g_i^{1d} - p_i^1))]x_j^d. \quad (3.14)$$

On the other hand, cost savings for deviated demand from other modes or activities are equal to:

$$\sum_{\substack{j=1 \\ j \neq i}}^{n} c_j^- x_j^d,$$

$$(3.15)$$

where c_j^- is the marginal operating cost of the alternative j, net of taxes.

3. Finally, there is a change in resources because of the costs of the HSR, and we have to consider them net of taxes: C_i^{RS-}; I^-; C_i^{I-}. We must also consider accidents cost savings ($\mathcal{E}$).

In sum, adding the changes in WTP and resources following the multimodal or corridor analysis, we obtain the change in social welfare, which is equal to:

$$\Delta W = \Delta \text{Resources}^e + \sum_{\substack{j=1 \\ j \neq i}}^{n} \Delta \text{Resources}_j^d = v_i\left(t_i^0 - t_i^1\right)x_i^0 + c_i^{0-}x_i^0$$

$$+ \sum_{\substack{j=1 \\ j \neq i}}^{n} \frac{1}{2}[((g_j^0 - p_j) - (g_i^{1d} - p_i^1)) + ((g_i^{1d} - p_j) - (g_i^{1d} - p_i^1))]x_j^d$$

$$+ \sum_{\substack{j=1 \\ j \neq i}}^{n} c_j^- x_j^d - C_i^{RS-} - I^- - C_i^{I-} + \mathcal{E}. \qquad (3.16)$$

3.5.2 Cost–Benefit Analysis of a New High-Speed Line: An Illustration

To illustrate the methodology, let us now consider a new HSR line with total length of 400 km, connecting two cities, already served by the conventional rail, air, and road transport (with cars and regular bus services). Once the HSR line is in operation, the conventional train services close down. The HSR is vertically unbundled and operated by two different companies: the infrastructure manager and the railway operator.[5]

In this practical illustration, we assume that construction works takes place from year 1 to year 5 and that the new infrastructure starts operation in year 6 (with no residual value). The NPV$_S$ of this project corresponds to the beginning of year 1, with an evaluation horizon of 30 years (including the investment period) and using a social discount rate of 3%. All the benefits and costs (euros in real terms) are assumed to be realized at the end of each year. The project will be compared to a 'do-nothing' alternative, where the described corridor continues to be served by the four initial modes. Parameters are summarized in Table 3.8 in the Appendix. Additional assumptions about costs and demand parameters are described below.

Costs and Demand Parameters Assumptions

As described above, the cost of building and operating a new HSR line can be broken down into four main items (see Table 3.1). Firstly, infrastructure construction costs, which include planning and land expropriation, construction works (including materials and labour) and the construction of stations. Although these values depend on the specific characteristics of the project (orography, land value, etc.), for the purposes of the case in this section, we will assume an average construction cost of €30 million per kilometre. Thus, the total construction cost of the line (400 km) is €12 billion, evenly spread over the years of construction. We assume that the residual value of the infrastructure will be zero after 30 years due to technical obsolescence.

Infrastructure operation and maintenance costs are also estimated at an average value of €100,000 per km once the line is in operation. It includes several components (electrification, signalling systems, track characteristics, weather conditions, etc.) which may vary according to the intensity of operations and the characteristics of

---

[5] For other case studies see, for example, de Rus (2012) and EIB (2023).

Table 3.1 Cost parameters assumptions

Parameter	Value	Unit
Infrastructure construction costs	30,000,000	€/km
Infrastructure operation and maintenance costs	100,000	€/km
Rolling stock acquisition costs	30,000,000	€/train
Rolling stock operation and maintenance costs	10,000,000	€/train

the rolling stock. Rolling stock acquisition costs are €30 million for a single-model 350-seat train with an economic life of 30 years.[6] Finally, rolling stock operation and maintenance costs are estimated at €10 million per train per year (Campos et al., 2009).

A value added tax (VAT) rate of 20% is charged for all inputs except labour throughout the life span of the project.[7] The share of labour costs is 20% of investment, 50% of operation and maintenance of the infrastructure, and 100% of operation and maintenance of rolling stock. With respect to shadow prices for labour, we have considered an average value of one, since, on the one hand, there are involuntary unemployed workers with a social opportunity cost lower than the gross wage (due to a reduction of a negative externality on unemployment); and, on the other hand, there are workers diverted from the rest of the economy with a social opportunity cost greater than the gross wage.[8] Moreover, no income tax is considered, and labour costs in operation and maintenance of infrastructure and rolling stock are assumed to grow in real values according to a real income growth rate of 2%, with a cost-income elasticity equal to one. The shadow price of public funds is also one.

Regarding the demand parameters, we need an estimate of the total number of passenger-trips in the corridor and the modal split with the project in each of the years included in the evaluation horizon. This estimate can be approached in different ways. We assume an annual volume of 20 million passenger-trips for the first year of operation. We also assume that the demand grows at the same rate as real income (2% in our illustration), with a demand-income elasticity of one. Demand estimation for the HSR line is completed by assuming an exogenous modal split summarized in Table 3.2.

For each of the transport modes in this example, Table 3.3 summarizes travel times, disaggregated into three main components: in-vehicle time, waiting time, and access/egress time. We have assumed waiting times of 20 min (40 min for air transport), and access/egress times of 40 min (75 min for air transport). In the case of car users, we have assumed their waiting and access/egress times are zero. All these values will be used to calculate time savings. A correction factor of 1.5 should be added to waiting

---

[6] This single-model assumption is used for the sake of simplicity, as there are many alternative trainsets and, of course, different trainsets can be combined on the same HSR line.

[7] We consider that all non-labour costs are not diverted from other economic activities.

[8] It includes the lost value of the marginal productivity of labour in the private sector, taking into account, for example, indirect taxes or social security contribution paid by employers, as discussed in Chap. 2.

Table 3.2 Assumptions on HSR demand

HSR demand from	%
Conventional train	40
Air transport	25
Bus	10
Car	20
Generated demand	5

Table 3.3 Time cost per passenger-trip for each mode of transport

Mode	In-vehicle value of time (€/h)	In-vehicle time (h)	Waiting value of time (€/h)	Waiting time (h)	Access and egress value of time (€/h)	Access and egress time (h)
Conventional train	15	3.50	22.5	0.33	22.5	0.66
Air transport	30	1.00	45	0.66	45	1.25
Bus	10	4.25	15	0.33	15	0.66
Car	15	3.50	22.5	0	22.5	0
HSR	–	1.80	–	0.33	–	0.66

time and an access/egress values of time (with respect to in-vehicle value of time), as well as that the fact that these values grow over time according to the evolution of real income (in this case, with an elasticity of 0.7).

The basic assumptions of the model are completed by estimating prices and unit costs for each of the transport modes. For prices, we have assumed a single and exogenous (average) fare for air, bus, and rail services (Table 3.4). For all modes, unit (avoidable) costs are equal to prices net of taxes.

Finally, in order to determine the cost of rolling stock, the number of trains purchased each year and the size of the fleet must first be calculated. To do this, we first estimate the daily (one-way) demand (assuming 365 days of service operation) and then the number of required daily services (DS), given by:

Table 3.4 Prices per passenger-trip for each mode of transport

Mode	Prices (€) (VAT included)	VAT (%)
Conventional train	30	10
Air transport	60	10
Bus	25	10
Car	55	30
HSR	50	10

$$DS = \frac{\text{Daily demand}}{(\text{Seats})(\text{Load factor})},$$

where the (target) load factor (defined as the ratio of seats-km over passenger-km) is assumed at 70% (on average). Assuming a maximum of 16 h of operation per day, the (minimum) number of required trains to meet the demand results from:

$$\text{Trains} = \frac{2(t_{\text{HSR}} + t_{\text{ST}})}{16} \cdot DS,$$

where t_{HSR} represents the total travel time, and t_{ST} is the time between trains (e.g., 0.5 h). The number of trains obtained from the previous expression defines a basic service. Both contingency factors (to avoid disruption due to breakdowns or periodic maintenance) and the existence of technical and safety limits (e.g., a maximum number of kilometres per train per year) imply that the final number of trains could be higher. We assume that trainsets are purchased the year they are needed and that there is no need to replace them during the project.

3.5.3 The Equivalence of the Different Approaches

Using all the formulas in Sect. 3.5.1 and the assumptions in Sect. 3.5.2, we can finally numerically prove the equivalence of the different CBA approaches to evaluate changes in social welfare due to a transport project: the change in surpluses approach, the change in WTP and resources approach considering the unimodal or single graph analysis, and the change in WTP and resources approach considering the multimodal or the corridor analysis. In particular, Table 3.5 displays the changes in the surpluses approach, whereas Tables 3.6 and 3.7, respectively, provide the results for the change in WTP and resources approach, according to the unimodal or single graph and the multimodal or the corridor analysis, respectively.

In practice, the choice of method critically depends on data availability. The first one (see Table 3.5) has the advantage of providing more information on the disaggregation of benefits and costs, but it is more demanding in terms of required information. The method base in the change in WTP and resources approach considering the unimodal or single graph analysis (see Table 3.6) is quite intuitive if multimodal effects are significant, but it is more difficult for identifying winners and losers. Finally, the method base on the change in WTP and resources approach, according to the multimodal or the corridor analysis (see Table 3.7), seems the easiest way to conduct the CBA evaluation when the effects on the rest of the economy can be safely ignored. If the information allows it, it is worth using the three methods simultaneously to minimize the likelihood of errors in the evaluation process since the final NPV_S must be identical applying the three methods.

Table 3.5 CBA of the HSR line: change in surpluses (thousands of euros)

(1)	**HSR consumers' surplus**	**4,016,351**
(2)	Conventional train	1,894,582
(3)	Air transport	1,507,723
(4)	Bus	119,542
(5)	Car	341,052
(6)	Generated	153,453
(7)	**Capital owners' surplus**	**− 4,824,172**
(8)	*HSR operator*	*3,839,874*
(9)	Revenues	16,972,803
(10)	Conventional train	6,789,121
(11)	Air transport	4,243,201
(12)	Bus	1,697,280
(13)	Car	3,394,561
(14)	Generated	848,640
(15)	Costs	− 13,132,930
(16)	Rolling stock acquisition + operation and maintenance	− 10,084,753
(17)	Infrastructure charges	− 3,048,177
(18)	*Infrastructure manager*	*− 8,664,046*
(19)	Infrastructure charges	3,048,177
(20)	Investment	− 10,991,297
(21)	Infrastructure operation and maintenance costs	− 720,926
(22)	*Other transport modes or activities*	*0*
(23)	Revenues lost	− 13,866,849
(24)	Conventional train	− 4,073,473
(25)	Air transport	− 5,091,841
(26)	Bus	− 848,640
(27)	Car	− 3,159,553
(28)	Generated	− 693,342
(29)	Cost savings	13,866,849
(30)	Conventional train	4,073,473
(31)	Air transport	5,091,841
(32)	Bus	848,640
(33)	Car	3,159,553
(34)	Generated	693,342
(35)	**Taxpayers' surplus**	**1,383,637**
(36)	*Taxes from investment*	*1,465,506*

(continued)

Table 3.5 (continued)

(37)	*Taxes from infrastructure operation and maintenance costs*	*50,069*
(38)	*Taxes from rolling stock acquisition + operation and maintenance costs*	*222,635*
(39)	*Taxes from other modes or activities*	*− 2,051,854*
(40)	Conventional train	− 407,347
(41)	Air transport	− 509,184
(42)	Bus	− 84,864
(43)	Car	− 947,866
(44)	Generated	− 102,593
(45)	*HSR taxes*	*1,697,280*
(46)	Conventional train	678,912
(47)	Air transport	424,320
(48)	Bus	169,728
(49)	Car	339,456
(50)	Generated	84,864
(51)	**Surplus of the rest of society**	**2,352,246**
(52) = (1) + (7) + (35) + (51)	SOCIAL NET PRESENT VALUE	2,928,062

3.6 Additional Economic Benefits of High-Speed Rail[9]

In this chapter, the economic appraisal of the HSR rests on the direct benefits of HSR in terms of time savings, WTP of generated traffic, and avoidable costs in competing modes. The likelihood that these benefits are enough to compensate for the fixed construction costs and the operating costs depends on the initial conditions, the counterfactual. When there exists an upgraded conventional railway able to run above 150 km and convenient air services, it is more difficult to find a socially profitable investment based exclusively on time savings, generated traffic, and cost savings in the conventional modes unless the volume of demand is very high.

Are these the only benefits of HSR or are we leaving important additional benefits like economies of agglomeration or significant environmental externalities? In our stylized case, the benefits of the HSR project come from existing passengers using the conventional railway service, the deviated demand from other transport modes, and the induced demand after the reduction of the generalized cost of travel. Therefore, total time savings heavily depend on the initial conditions in the transport corridor. The potential direct benefits of HSR are expected to be significant when there are capacity constraints in the conventional rail network, roads, and airports.

---

[9] This section draws on de Rus (2021b).

Table 3.6 CBA of the HSR line: change in WTP and resources using the unimodal or single graph approach (thousands of euros)

(53)	Existing demand (conventional train)	8,955,268
(54) = (2) + (10) + (24) + (40) + (46)	Time savings	4,881,795
(55) = (30)	Cost savings	4,073,473
(56)	**Change in WTP**	**32,852,511**
(57)	Air transport	17,966,932
(58)	Bus	3,558,775
(59)	Car	8,791,744
(60)	Generated	2,535,059
(61)	**Time costs**	**− 19,528,691**
(62)	Air transport	− 11,791,688
(63)	Bus	− 1,572,225
(64)	Car	− 4,716,675
(65)	Generated	− 1,448,102
(66)	**Taxes in other modes or activities**	**− 1,644,507**
(67)	Air transport	− 509,184
(68)	Bus	− 84,864
(69)	Car	− 947,866
(70)	Generated	− 102,593
(71) = (51)	**Accident costs**	**2,352,246**
(72) = (20) + (36)	**Investment**	**− 9,525,791**
(73) = (21) + (37)	**Infrastructure operation and maintenance costs**	**− 670,857**
(74)	**Rolling stock acquisition costs**	**− 1,113,174**
(75)	**Rolling stock operation and maintenance costs**	**− 8,748,943**
(76) = (53) + (56) + (61) + (66) + (71) + (72) + (73) + (74) + (75)	SOCIAL NET PRESENT VALUE	2,928,062

Other equivalences between the data provided in Tables 3.5 and 3.6 are as follows:

(57) + (62) + (67) = (3) + (11) + (25) + (31) + (41) + (47)
(58) + (63) + (68) = (4) + (12) + (26) + (32) + (42) + (48)
(59) + (64) + (69) = (5) + (13) + (27) + (33) + (43) + (49)
(60) + (65) + (70) = (6) + (14) + (28) + (34) + (44) + (50)
(74) + (75) = (16) + (38)

Table 3.7 CBA of the HSR line: change in WTP and resources using the multimodal or corridor approach (thousands of euros)

(77)	Time savings	6,767,732
(78) = (2) + (10) + (24) + (40) + (46) = (54)	Conventional train	4,881,795
(79) = (3) + (11) + (25) + (41) + (47)	Air transport	574,219
(80) = (4) + (12) + (26) + (42) + (48)	Bus	1,053,046
(81) = (5) + (13) + (27) + (43) + (49)	Car	− 32,350
(82) = (6) + (14) + (28) + (44) + (50)	Generated	**291,022**
(83)	**Cost savings**	13,866,849
(84) = (30) = (55)	Conventional train	4,073,473
(85) = (31)	Air transport	5,091,841
(86) = (32)	Bus	848,640
(87) = (33)	Car	3,159,553
(88) = (34)	Generated	693,342
(89) = (51) = (71)	**Accident costs**	**2,352,246**
(90) = (20) + (36) = (72)	**Investment**	**− 9,525,791**
(91) = (21) + (37) = (73)	**Infrastructure operation and maintenance costs**	**− 670,857**
(92)	**Rolling stock acquisition costs**	**− 1,113,174**
(93)	**Rolling stock operation and maintenance costs**	**− 8,748,943**
(94) = (77) + (83) + (89) + (90) + (91) + (92) + (93)	SOCIAL NET PRESENT VALUE	2,928,062

Other equivalences between the data provided in Tables 3.5, 3.6 and 3.7 is as follows:

(79) + (85) = (57) + (62) + (67)
(80) + (86) = (58) + (63) + (68)
(81) + (87) = (59) + (64) + (69)
(82) + (88) = (60) + (65) + (70)
(92) + (93) = (74) + (75) = (16) + (38)

Even changes in modal split should be carefully examined. The critical issue is the existence of distortions in the secondary markets, i.e., price is higher or lower than marginal social cost in the alternative mode of transport. When price is below marginal social cost in the alternative transport mode, the deviation of demand to the new transport mode (assuming price is at least equal to marginal social cost in the new mode) adds new benefits to the project, like the reduction of suboptimal congestion or pollution. On the contrary, if the externality is positive in the alternative mode, the deviation of traffic to HSR reduces the benefit of the project, like in the case of the reduction of service frequencies both in air and bus transport following the introduction of a new HSR line.

A sound economic appraisal should consider the relevant alternatives to solve the transport problem at stake. One of these alternatives is the introduction of optimal pricing. The reduction of road congestion and airport delays is a positive effect of HSR with suboptimal pricing. Nevertheless, when road and airport congestion charges internalize the externalities, there are not indirect benefits from shifting traffic between modes.

It is not uncommon that promoters of HSR projects emphasize the indirect effects, wider economic benefits (WEBs) and regional development instead of the direct effects of the new rail service. Although transport investments produce benefits beyond the direct benefits it is unclear whether these additional benefits are really additional or double counting. Moreover, some indirect benefits are in many cases common to any other infrastructure project, so even if the benefits increase the social return on the investment in transport, they do not affect the net welfare effect of the project under evaluation.

Recent research suggests that agglomeration benefits in sectors such as financial services may be greater than in manufacturing. This is relevant to the urban commuting case and may be applicable to some HSR services (e.g., networks which links a set of major financial centres and may be used as a form of weekly commuting).

Regarding environmental benefits, HSR infrastructure is one of the feasible alternatives to deal with transport capacity problems in intercity transport. The analysis of the life cycle of HSR investment is needed to determine the final effect on the environment. The net effect depends on the volume of demand deviated from less environmentally friendly transport modes and whether the demand is high enough to compensate the negative externalities during construction, and also other neglected but significant negative externalities like the barrier effect, noise, and visual intrusion.

Another alleged benefit of HSR investment is social cohesion and the correction of territorial imbalances. The evidence is not supporting this hypothesis. HSR lines tend to favour central locations, so it may be not beneficial for the periphery. When HSR investment allows the expansion of markets and the exploitation of economies of scale and agglomeration, it may encourage the location of jobs in major urban centres where there are agglomeration benefits.

3.7 Conclusions

HSR is a transport technology that provides fast, comfortable, reliable, and safe rail services to attend medium distance intercity mobility. It requires a high volume of demand, with a high WTP for time savings and other utility components of this service, to compensate for the high sunk costs of construction of this infrastructure. It is worth to remember that the development of HSR networks is not, in general, determined by market forces, and therefore, the economic appraisal of projects is crucial to inform governments of the expected net social benefits of new lines.

This chapter provides two alternative and equivalent approaches for the evaluation of the direct benefits and some indirect effects of HSR projects. There may exist additional benefits, particularly the WEBs derived from the response of workers and firms to changes in proximity. These effects as well as the environmental effects during the life span of the projects should be included when there is evidence of their relevance. Additionally, positive effects of concentration have its downside (e.g., congestion in the centre and productivity losses in the periphery) and should be included in the evaluation.

We have used arbitrary prices in the evaluation of the stylized HSR investment to illustrate the methodology. Pricing is crucial in the evaluation of actual projects. Different pricing policies may significantly affect the social profitability of projects (see Chap. 4). A sound economic appraisal of projects requires an explicit consideration of pricing policy both for the HSR itself and competing transport modes. The economic evaluation of capacity expansion may produce quite different results depending on the pricing scheme in the counterfactual.

Appendix

Table 3.8 Values and assumptions for the numerical illustration

Length: 400 km
First year of construction: 1
Last year of construction: 5
First year of operation: 6
Project life: 30 years
Social discount rate: 3%
Income annual growth rate: 2%
Benefits and costs are located at the end of the year and expressed in real terms
Shadow price of labour: 1
Shadow price of public funds: 1
Elasticity of labour costs with respect to income: 1
Elasticity of the demand with respect to income: 1
Elasticity of the value of time with respect to income: 0.7
Elasticity of the accident costs with respect to income: 1
VAT of investment, operation and maintenance costs, and acquisition of rolling stock: 20%
Investment cost per km: €30,000,000
Investment costs are linearly distributed during the construction period

(continued)

Table 3.8 (continued)

Labour share in investment costs: 20%

Labour share in infrastructure maintenance costs: 50%

Labour share in operation and maintenance costs of the rolling stock: 100%

Residual value of the infrastructure: 0

Operation and maintenance cost of the infrastructure: 100,000 €/km

Operation and maintenance cost per train: €10,000,000 per year

Infrastructure charges: 10€ train-km

Acquisition of rolling stock: €30,000,000 per train

Average capacity per train: 350 seats

Train life: 30 years

For the number of trains purchased each year, it is assumed: contingency factor: 1.15; no maximum number of kilometres per year; headway: 0.5 h; no seasonality; load factor: 0.7; hours of operation: 16

Average avoidable cost in other activities are equal to their prices net of taxes

First year demand: 20,000,000 passenger-trips

Demand is computed considering the number of passenger-trips of the previous year, the annual growth rate of the income and elasticity of the demand with respect to income

Modal split: conventional train 40%; air transport 25%; bus 10%; car 20%; generated 5%

The values for generated demand are obtained according to the distribution of deviated demand

Accident costs (€ per passenger-km): conventional train 0.005; air transport 0.0001; bus 0.01; car 0.045; HSR 0.001

Chapter 4
The Consequences of Pricing Policies on Investment Decisions

4.1 Introduction

As already explained in Chap. 2 (Sect. 2.4), prices affect demand and, thus, the social surplus of investment projects. Therefore, before evaluating a transport infrastructure and its alternatives, we need to know the associated charging scheme. In this chapter, we analyse the effects of different charging schemes when deciding between alternative transport modes.[1] Pricing decisions affect differently the social welfare of alternative transport modes. Thus, when comparing different transport alternatives, a particular charging scheme may favour the creation of a particular transport infrastructure network, leading to long-term equilibria that might not be optimal under other initial charging schemes.

Different charging schemes may be applied in practice. The first possibility consists of charging according to short-run marginal costs. The second one consists of charging mark-ups over short-run marginal costs to cover full costs (both operation and construction costs).

In Sect. 4.2, we develop a theoretical model in which users demand services in two transport modes: air transport and railway. Given users' preferences, the regulator must decide whether to invest in air transport, railway, both transport modes, or postpone the investment, using two possible charging schemes: either charging according to short-run marginal costs or charging mark-ups over short-run marginal costs. In this context, we show that the regulator may favour the construction of the rail infrastructure by choosing a charging scheme based on short-run marginal costs. The game consists of three stages and it is solved by backward induction in Sect. 4.3. Section 4.4 includes an empirical illustration of the model. In Sect. 4.5, we summarize the main conclusions of this chapter.

---

[1] This chapter is based on de Rus & Socorro (2019).

4.2　The Model

Let us consider an economy composed of an oligopolistic transport sector and a competitive (numeraire) sector summarizing the rest of the economy. The transport sector contains two transport modes: rail and air transport. Transport infrastructures are public and used by private operators. In particular, we assume that the rail infrastructure is used by a private operator while two private airlines operate in airports.

Denote by x_1, x_2, x_r the quantity offered and demanded on a certain route by airline 1, airline 2 and the rail operator, respectively. On this corridor, there exist N identical users with a utility function separable and linear in the numeraire good, m: $U(x_1, x_2, x_r) + m$. Therefore, there are no income effects on the transport sector, and we can perform a partial equilibrium analysis.

The utility function of the representative user in the transport sector, $U(x_1, x_2, x_r)$, is assumed to be quadratic and strictly concave:

$$U(x_1, x_2, x_r) = u_a x_1 + u_a x_2 + u_r x_r - \frac{1}{2}\left(x_1^2 + x_2^2 + x_r^2 + 2\gamma x_1 x_2 + 2\delta x_1 x_r + 2\delta x_2 x_r\right),$$

$$(4.1)$$

where u_a and u_r are positive parameters that measure user's preferences for each transport mode, γ represents the degree of product differentiation between airlines, and δ represents the degree of product differentiation between airlines and the railway. We assume that passengers consider that airlines are substitutes but exhibit brand loyalty to particular carriers, i.e., airlines compete in differentiated products.[2] Therefore, $\gamma \in [0, 1)$. When the parameter γ is zero, airlines are independent. As γ tends to one, airlines are considered better substitutes.

Moreover, passengers consider the railway and airlines as substitutes. Therefore, we assume that $\delta \in [0, 1)$. As δ tends to one, rail and airlines are considered better substitutes. The parameter δ is equal to zero when passengers consider rail and airlines as independent goods. However, we assume that $\gamma > \delta$, which implies that passengers consider that the degree of substitutability between one airline and the train is lower than the degree of substitutability between one airline and the other one.

Generalized price is defined as the sum of the ticket-price, p_j with $j = 1, 2, r$, and the monetary value of time and/or any disutility component associated with the specific transport mode, θ_a and θ_r, which includes access, egress, waiting and in-vehicle time, discomfort, etc. Thus, the representative user solves:

$$\max_{x_1, x_2, x_r} U(x_1, x_2, x_r) - (p_1 + \theta_a)x_1 - (p_2 + \theta_a)x_2 - (p_r + \theta_r)x_r. \qquad (4.2)$$

---

[2] Product differentiation between airlines may be due to several reasons such as brand loyalty or the existence of frequent flier programs.

The above maximization program can be rewritten as:

$$\max_{x_1, x_2, x_r} \alpha x_1 + \alpha x_2 + \beta x_r - \frac{1}{2}\left(x_1^2 + x_2^2 + x_r^2 + 2\gamma x_1 x_2 + 2\delta x_1 x_r + 2\delta x_2 x_r\right)$$
$$- p_1 x_1 - p_2 x_2 - p_r x_r, \tag{4.3}$$

where $\alpha = u_a - \theta_a$ and $\beta = u_r - \theta_r$ denote the maximum (net of all except ticket-price) willingness to pay (WTP) for travelling by air or by rail, respectively.

Let us denote by μ_a and μ_r the access price charged to airlines and the railway operator, respectively, and by c_a and c_r the constant marginal operating costs of each transport operator. Denote by c_A and c_R transport infrastructure marginal operating and maintaining costs.

We consider two possible access charging policies commonly used in practice. The first one consists of charging for the use of transport infrastructures according to short-run marginal costs, i.e., $\mu_a = c_A$ and $\mu_r = c_R$, respectively. The second one consists of charging a mark-up over short-run marginal costs in order to cover part of the construction costs, i.e.: $\mu_a = c_A + A$ and $\mu_r = c_R + R$, where A and R represent mark-ups over short-run marginal costs. Let us denote by I_A and I_R the construction costs of an airport and the railway infrastructure, respectively, with $I_A < I_R$, and by r_S the social discount rate.

For every segment connecting two regions, the timing of the game is the following: In the first stage, the regulator decides the access pricing scheme: either charging transport operators according to short-run marginal costs or, alternatively, charging a mark-up over short-run marginal costs. Then, given the charging scheme, the regulator decides whether to construct (or not) airports, the rail infrastructure or both transport infrastructures to connect regions. In the second stage, given the charging scheme and the transport infrastructures that were built, private operators pay access prices and decide the ticket-price to be charged to final users. Finally, in the third stage, given the ticket-price, each user demands a certain number of trips in those transport modes for which transport infrastructures were built. The game is solved by backward induction.

4.3 Optimal Transport Infrastructures

4.3.1 Stage 3: Users' Demand

In the last stage of the game, given the ticket-price, the representative user demands a certain number of trips in those transport modes for which transport infrastructures were built.

If only airports were constructed, the representative user demands air transport trips to airline 1 and airline 2. Let us denote by x_1^a and x_2^a the quantity demanded by the representative user to airline 1 and airline 2 if only airports were constructed. Then, the representative user solves:

$$\max_{x_1,x_2} \alpha x_1 + \alpha x_2 - \frac{1}{2}\left(x_1^2 + x_2^2 + 2\gamma x_1 x_2\right) - p_1 x_1 - p_2 x_2, \qquad (4.4)$$

which leads to the following linear demand functions for airlines:

$$x_1^a = a - bp_1 + dp_2; \quad x_2^a = a - bp_2 + dp_1, \qquad (4.5)$$

where $a = \frac{\alpha(1-\gamma)}{1-\gamma^2}$, $b = \frac{1}{1-\gamma^2}$, $d = \frac{\gamma}{1-\gamma^2}$.

On the other hand, if only the rail infrastructure was constructed, the representative user demands rail transport trips to the rail operator. Let us denote by x_r^r the number of rail trips demanded by the representative user if only the rail infrastructure was constructed. Then, the representative user solves:

$$\max_{x_r} \beta x_r - \frac{1}{2}x_r^2 - p_r x_r, \qquad (4.6)$$

which leads to the following linear demand functions for the rail operator:

$$x_r^r = \beta - p_r. \qquad (4.7)$$

Finally, if both airports and the rail infrastructure were constructed, the representative user demand both air transport and train trips. Let us denote by x_1^{a+r}, x_2^{a+r}, x_r^{a+r} the quantity demanded by the representative user to airline 1, airline 2 and the rail operator when both airports and the rail infrastructure were constructed. In this case, the representative user solves the maximization program given by expression (4.3), leading to the following linear demand functions for airlines and the rail operator:

$$x_1^{a+r} = a_a - b_a p_1 + d_a p_2 + d_r p_r,$$
$$x_2^{a+r} = a_a - b_a p_2 + d_a p_1 + d_r p_r,$$
$$x_t^{a+r} = a_r - b_r p_r + d_r p_1 + d_r p_2, \qquad (4.8)$$

where $a_a = \frac{(\alpha-\beta\delta)}{1+\gamma-2\delta^2}$, $b_a = \frac{(1-\delta^2)}{(1-\gamma)(1+\gamma-2\delta^2)}$, $d_a = \frac{(\gamma-\delta^2)}{(1-\gamma)(1+\gamma-2\delta^2)}$, $a_r = \frac{\beta(1+\gamma)-2\alpha\delta}{1+\gamma-2\delta^2}$,

$b_r = \frac{1+\gamma}{1+\gamma-2\delta^2}$, $d_r = \frac{\delta}{1+\gamma-2\delta^2}$.

We assume that $\alpha - \beta\delta > 0$, and $\beta(1 + \gamma) - 2\alpha\delta > 0$. Given these assumptions, all the parameters $(a_a, b_a, d_a, a_r, b_r, yd_r)$ are strictly positive.

4.3.2 Stage 2: Optimal Ticket-Prices

In the second stage, given the charging scheme and the transport infrastructures that were built, private operators pay access prices (μ_a and/or μ_r) and decide the ticket-price to be charged to final users.

If only airports were constructed, airlines would compete in differentiated products, solving the following maximization program:

$$\max_{p_j} \pi_j^a = (p_j - c_a - \mu_a)x_j^a, \tag{4.9}$$

where π_j^a represents the profit for airline j and x_j^a represents user's demands given by expression (4.5), with $j = 1, 2$. Optimal ticket-prices for airline 1 and airline 2 if only airports were constructed are then given by:

$$p_1^a = p_2^a = \frac{a + b\mu_a + bc_a}{2b - d}. \tag{4.10}$$

If only the rail infrastructure were constructed, the rail operator would solves the following maximization program:

$$\max_{p_r} \pi_r^r = (p_r - c_r - \mu_r)x_r^r, \tag{4.11}$$

where π_r^r represents rail operator profits and x_r^r is the representative user's demand given by expression (4.7). Optimal ticket-price for the rail service when only the rail infrastructure was constructed is given by:

$$p_r^r = \frac{\beta + \mu_r + c_r}{2}. \tag{4.12}$$

If both airports and the rail infrastructure were constructed, airlines and the rail operator would compete in differentiated products. On the one hand, airline j solves:

$$\max_{p_j} \pi_j^{a+r} = (p_j - c_a - \mu_a)x_j^{a+r}, \tag{4.13}$$

where x_j^{a+r} is given by expression (4.8), with $j = 1, 2$.

On the other hand, the rail operator solves:

$$\max_{p_r} \pi_r^{a+r} = (p_r - c_r - \mu_r)x_r^{a+r}, \tag{4.14}$$

where x_r^{a+r} is given by expression (4.8).

Solving airlines and rail operator maximization programs, we obtain the following optimal ticket-prices for air transport and rail trips if both airports and the rail infrastructure were constructed:

$$p_1^{a+r} = p_2^{a+r} = \frac{1}{4b_a b_r - 2d_r^2 - 2d_a b_r}$$
$$(2a_a b_r + a_r b_r + 2\mu_a b_a b_r + \mu_r b_r d_r + 2b_a c_a b_r + b_r c_r d_r),$$

$$p_r^{a+r} = \frac{1}{4b_a b_r - 2d_r^2 - 2d_a b_r}(2b_a a_r + 2a_a d_r - d_a a_r + 2\mu_a b_a d_r + 2\mu_r b_a b_r$$
$$- \mu_r d_a b_r + 2b_a c_a d_r + 2b_a b_r c_r - d_a b_r c_r). \tag{4.15}$$

4.3.3   Stage 1: Optimal Transport Infrastructure Provision

In the first stage, the regulator decides the access pricing scheme: either charging transport operators according to short-run marginal costs or, alternatively, charging a mark-up over short-run marginal costs. Then, given the charging scheme, the regulator decides whether to construct (or not) airports, the rail infrastructure or both transport infrastructures.

In order to optimally choose which transport infrastructure should be constructed, the regulator should compare the social welfare associated with each possible alternative. Social welfare for each alternative is defined as the social net present value (NPV$_s$) of each project, obtained as the present value of consumers' surplus, transport operators' surplus, and the surplus due to the use of transport infrastructures, minus the investment cost. Let us define $\widehat{W}$ as the first year social welfare of a project (obtained as the sum of the first year consumers' surplus, transport operators' surplus, and the surplus due to the use of transport infrastructures) minus the investment opportunity cost.[3] With this definition and considering infinite-life infrastructure projects, a positive first year social welfare implies, on the one hand, that it is optimal to construct the transport infrastructure today instead of postponing the investment one period and, on the other hand, that the NPV$_s$ associated with such an infrastructure is positive (see the Appendix for a formal explanation of this interpretation).

The regulator has four possible alternatives:

Alternative 0: To construct no infrastructure today and postpone the investment till the number of users in the economy, N, is higher.

Alternative 1: To construct just the air transport infrastructure. The first year social welfare associated with this alternative, $\widehat{W}_a$, is given by:

---

[3] We consider no economic cost of public funds; i.e., the shadow price of public funds is equal 1.

$$\widehat{W}_a = N\big[U\big(x_1^a, x_2^a\big) - \big(p_1^a + \theta_a\big)x_1^a - \big(p_2^a + \theta_a\big)x_2^a + \pi_1^a$$
$$+ \pi_2^a + (\mu_a - c_A)\big(x_1^a + x_2^a\big)\big] - r_S I_A, \tag{4.16}$$

which can be rewritten as:

$$\widehat{W}_a = N\left[\alpha x_1^a + \alpha x_2^a - \frac{1}{2}\Big[\big(x_1^a\big)^2 + \big(x_2^a\big)^2 + 2\gamma x_1^a x_2^a\Big] - (c_A + c_a)\big(x_1^a + x_2^a\big)\right] - r_S I_A, \tag{4.17}$$

where x_1^a and x_2^a represent user's demands given by expression (4.5) when considering optimal ticket-prices given by expression (4.10).

Alternative 2: To construct just the rail infrastructure. The first year social welfare associated with this alternative, $\widehat{W}_r$, is given by:

$$\widehat{W}_r = N\big[U\big(x_r^r\big) - \big(p_r^r + \theta_r\big)x_r^r + \pi_r^r + (\mu_r - c_R)x_r^r\big] - r_S I_R, \tag{4.18}$$

which can be rewritten as:

$$\widehat{W}_r = N\left[\beta x_r^r - \frac{1}{2}\big(x_r^r\big)^2 - (c_R + c_r)x_r^r\right] - r_S I_R, \tag{4.19}$$

where x_r^r represents user's demand given by expression (4.7) when considering the optimal ticket-price given by expression (4.12).

Alternative 3: To construct both the air and rail infrastructures. The first year social welfare associated with this alternative, $\widehat{W}_{a+r}$, is given by:

$$\widehat{W}_{a+r} = N\big[U\big(x_1^{a+r}, x_2^{a+r}, x_r^{a+r}\big) - \big(p_1^{a+r} + \theta_a\big)x_1^{a+r} - \big(p_2^{a+r} + \theta_a\big)x_2^{a+r}$$
$$- \big(p_r^{a+r} + \theta_r\big)x_r^{a+r} + \pi_1^{a+r} + \pi_2^{a+r} + \pi_r^{a+r} + (\mu_a - c_A)\big(x_1^{a+r} + x_2^{a+r}\big)$$
$$+ (\mu_r - c_R)x_r^{a+r}\big] - r_S I_A - r_S I_R, \tag{4.20}$$

which can be rewritten as:

$$\widehat{W}_{a+r} = N\left[\alpha x_1^{a+r} + \alpha x_2^{a+r} + \beta x_r^{a+r} - \frac{1}{2}\Big[\big(x_1^{a+r}\big)^2 + \big(x_2^{a+r}\big)^2 + \big(x_r^{a+r}\big)^2\right.$$
$$+ 2\gamma x_1^{a+r} x_2^{a+r} + 2\delta x_1^{a+r} x_r^{a+r} + 2\delta x_2^{a+r} x_r^{a+r}\Big]$$
$$\left. - (c_A + c_a)\big(x_1^a + x_2^a\big) - (c_R + c_r)x_r^{a+r}\right] - r_S I_A - r_S I_R, \tag{4.21}$$

where x_1^{a+r}, x_2^{a+r} and x_r^{a+r} represent user's demands given by expression (4.8) when considering optimal ticket-prices given by expression (4.15).

Notice that the first year social welfare function associated with each alternative is linear and strictly increasing in N (the number of users). When $N = 0$, the first year social welfare associated with each alternative is just the investment opportunity cost used to construct today the transport infrastructure. The slope of the first year social welfare function associated with each alternative depends on the parameters of the model and the charging scheme used by the regulator, μ_a and μ_r. The regulator may charge the use of transport infrastructures according to short-run marginal costs, i.e., $\mu_a = c_A$ and $\mu_r = c_R$; or charge a mark-up over short-run marginal costs, i.e., $\mu_a = c_A + A$ and $\mu_r = c_R + R$ where A and R represent mark-ups.

Since the slope of the first year social welfare functions depends on the parameters of the model and the charging scheme chosen by the regulator, we may have different situations.

Figure 4.1 depicts a situation in which if the number of users in the economy is lower than N_1 the optimal alternative is *alternative 0*, i.e., construct no infrastructure today and wait till the number of users increases. However, if the number of users in the economy is between N_1 and N_2 the regulator should optimally choose *alternative 1*, i.e., to construct just the air transport infrastructure. When the number of users is between N_2 and N_3, the optimal decision is to choose *alternative 2*, i.e., to construct just the rail infrastructure. Finally, if the number of users is high enough, in particular, higher than N_3, the optimal alternative is *alternative 3*, i.e., to construct both the air and rail infrastructures.

Figure 4.2 illustrates a situation in which, given the slope of the first year social welfare functions associated with each alternative, if the number of users in the economy is lower than N_1 the best alternative for the regulator is *alternative 0*, i.e., to construct no infrastructure today and postpone the investment till the number of

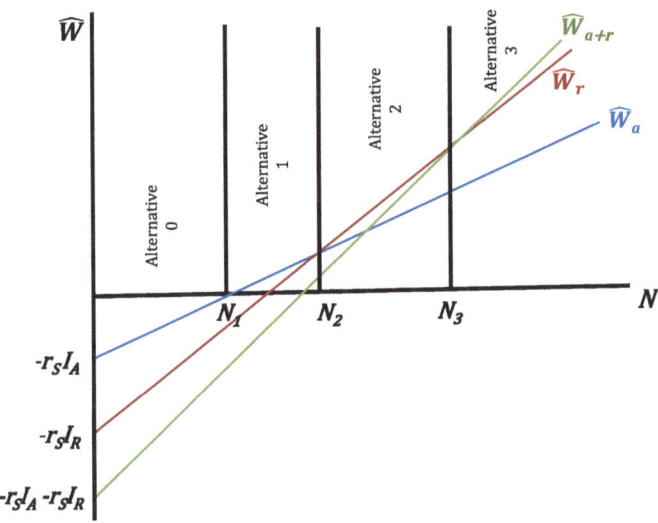

Fig. 4.1 Optimal alternative depending on the number of users in the economy: Example 1

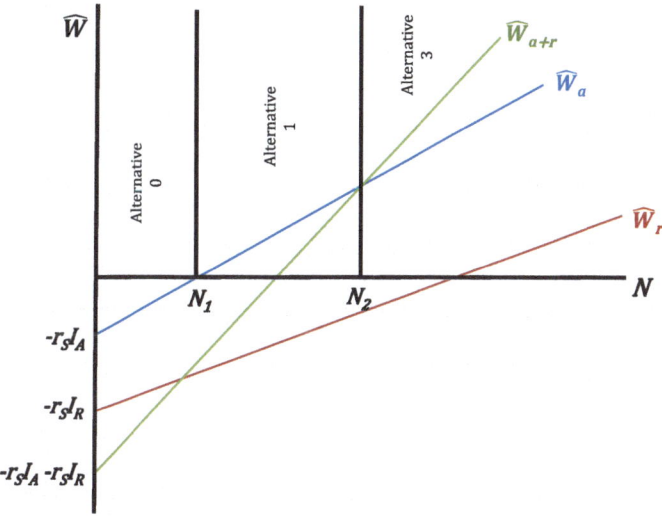

Fig. 4.2 Optimal alternative depending on the number of users in the economy: Example 2

users is higher. However, if the number of users is between N_1 and N_2, the regulator should optimally choose *alternative 1*, i.e., to construct just the air infrastructure. If the number of users in the economy is high enough, in particular, higher than N_2, the optimal choice is *alternative 3*, i.e., to construct both the air and rail infrastructures. In this example, *alternative 2* (to construct just the rail infrastructure) is never optimal.

Figure 4.3 represents another situation in which, given the slope of the first year social welfare functions associated with each alternative, if the number of users in the economy is lower than N_1 the best alternative for the regulator is *alternative 0*, i.e., to construct no infrastructure today and postpone the investment till the number of users is higher. However, if the number of users is between N_1 and N_2, the regulator should optimally choose *alternative 2*, i.e., to construct just the rail infrastructure. If the number of users in the economy is high enough, in particular, higher than N_2, the optimal choice is *alternative 3*, i.e., to construct both the air and rail infrastructures. In this example, *alternative 1* (to construct just the air infrastructure) is never optimal.

The slope of the first year social welfare function associated with each alternative depends on the charging scheme chosen by the regulator and the parameters of the model (such as user's preferences for each transport mode, the degree of product differentiation between airlines and between air and rail transport, marginal operating costs for transport operators, and marginal operating and maintaining costs of transport infrastructures). The regulator cannot influence the parameters of the model but can affect the slope of the first year social welfare function through the charging scheme. The higher the access prices are, the flatter is the first year social welfare function associated with each alternative. Thus, if the regulator moves from a charging scheme based on short-run marginal costs to a charging scheme based on

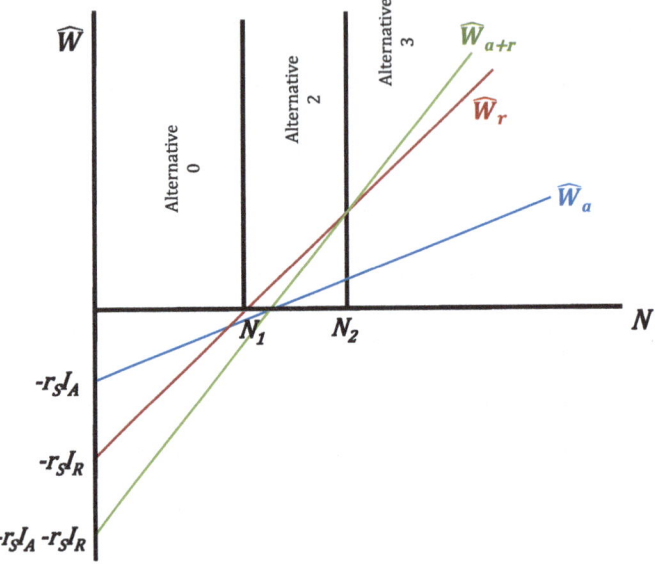

Fig. 4.3 Optimal alternative depending on the number of users in the economy: Example 3

mark-ups over short-run marginal costs, the first year social welfare function associated with each alternative becomes flatter. Since $I_A < I_R$ and mark-ups aim to cover part of the construction costs of each transport infrastructure, $A < R$, and the effect on the slope of the first year social welfare function when moving from one charging scheme to the other is higher in the case of the rail infrastructure. This is illustrated in Fig. 4.4, where dashed lines represent first year social welfare functions when using a charging scheme according to short-run marginal costs and solid lines represent first year social welfare functions when moving to a charging scheme based on mark-ups over short-run marginal costs.

Figure 4.4 depicts a situation in which if the regulator uses a charging scheme based on short-run marginal costs and the number of users in the economy is lower than N_1, the best alternative for the regulator is *alternative 0*, i.e., to construct no infrastructure today and postpone the investment till the number of users is higher. However, if the number of users is between N_1 and N_2, the regulator should optimally choose *alternative 1*, i.e., to construct just the air infrastructure. If the number of users in the economy is high enough, in particular, higher than N_2, the optimal choice is *alternative 3*, i.e., to construct both the air and rail infrastructures. In this example, *alternative 2* (to construct just the rail infrastructure) is never optimal. When the regulator decides to charge transport infrastructure access using mark-ups over short-run marginal costs, critical thresholds for the number of users move from N_1 to N_1' and from N_2 to N_2'. Thus, we need more users for *alternative 1* to be optimal and proportionally much more users for *alternative 3* to be chosen.

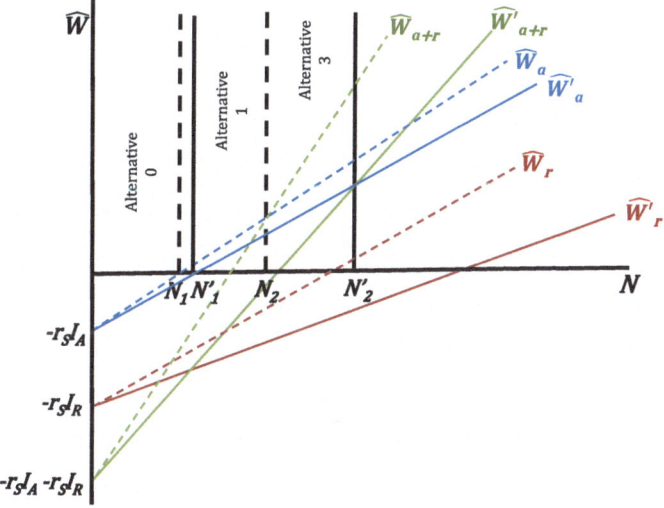

Fig. 4.4 Optimal alternative depending on the charging scheme

4.4 Empirical Illustration

In order to illustrate how the charging scheme chosen by the regulator affects demand thresholds and, hence, the optimality of transport infrastructures let us consider the following empirical illustration. Suppose a corridor within the range of 600–650 km length. For such a route, we will consider two possible transport modes: air transport and high-speed rail (HSR).

Taking into account real construction data in routes with similar length (see Chap. 3), in this empirical illustration we will consider that the total investment required to construct the rail infrastructure is $I_R = €10$ billion. On the other hand, the investment in airport capacity is assumed to be €1 billion. However, since for operating the route two airports are needed, we assume that $I_A = €2$ billion. The social discount rate is assumed to be 5%, i.e., $r_S = 0.05$.

Moreover, we assume that, on the one hand, the marginal operating and maintaining cost of the air transport infrastructure is $c_A = €10$. On the other hand, we assume that the marginal operating and maintaining cost of the HSR infrastructure is $c_R = €20$. Regarding transport services, we consider that the marginal operating cost for airlines and HSR is $c_a = €40$ and $c_r = €30$, respectively.

Regarding users' preferences, we assume that they consider airlines and the HSR as good substitutes, i.e., $\delta = 0.7$. The degree of product differentiation between airlines is also assumed to be low, i.e., $\gamma = 0.8$. Finally, we assume that the maximum (net of all except ticket-price) WTP for travelling by air or by HSR is $\alpha = 160$ and $\beta = 180$.

Table 4.1 shows optimal prices, quantities, and first year social welfare functions associated with each alternative when the regulator charges the use of transport infrastructures according to short-run marginal costs, i.e., $\mu_a = c_A$ and $\mu_r = c_R$. It also shows the minimum number of users, N, required for each alternative to be optimal. Since each user demands a certain number of trips, we also compute the minimum number of trips requires for each alternative to be optimal. Thus, if the number of users in the economy is lower than 15,301 (or, in other words, the total number of trips is lower than 1,558,407 trips), the best alternative for the regulator is *alternative 0*, i.e., to construct no infrastructure today and postpone the investment till the number of users is higher. However, if the number of users is between 15,301 and 258,730, the regulator should optimally choose *alternative 1*, i.e., to construct just the air infrastructure. If the number of users in the economy is higher than 258,730 (or, in other words, the total number of trips is higher than 31,914,345 trips, 16,964,926 trips by air, and 14,949,419 trips by HSR), the optimal choice is *alternative 3*, i.e., to construct both the air and rail infrastructures. *Alternative 2* (to construct just the HSR infrastructure) is never optimal (see Fig. 4.2 in Sect. 4.3).

If the regulator decides to charge for the use of transport infrastructure through mark-ups over short-run marginal costs, then $\mu_a = c_A + A$ and $\mu_r = c_R + R$. Since $I_A < I_R$ and mark-ups aim to cover part of the construction costs of each transport infrastructure, $A < R$. In our empirical illustration we assume that $A = 2$ and $R = 30$.

Table 4.2 shows optimal prices, quantities, first year social welfare functions, and users' thresholds associated with each alternative when the regulator charges the use of transport infrastructures according to mark-ups over short-run marginal costs. As shown in Table 4.2, if the number of users in the economy is lower than 15,385 (or, in other words, the total number of trips is lower than 1,538,500 trips), the best alternative for the regulator is *alternative 0*, i.e., to construct no infrastructure

Table 4.1 Prices and thresholds when charging according to short-run marginal costs

	Alternative 1: Air transport	*Alternative 2*: HSR	*Alternative 3*: Air transport and HSR
Access prices (€)	$\mu_a = 10$	$\mu_r = 20$	$\mu_a = 10;\ \mu_r = 20$
Ticket-prices (€)	$p_1 = p_2 = 68.33$	$p_r = 115$	$p_1 = p_2 = 60.54;$ $p_r = 76.32$
Number of trips per user	$x_1 + x_2 = 102$	$x_r = 65$	$x_1 + x_2 = 66;$ $x_r = 58$
First year social welfare functions	$\widehat{W}_a =$ 6535.5N-100,000,000	$\widehat{W}_r =$ 6337.5N-500,000,000	$\widehat{W}_{a+r} =$ 8468N-600,000,000
Minimum number of users	$N = 15,301$	–	$N = 258,730$
Minimum number of trips	$N(x_1 + x_2) =$ 1,558,407 Total: 1,558,407	–	$N(x_1 + x_2) =$ 16,964,926 $N x_r = 14,949,419$ Total: 31,914,345

Table 4.2 Prices and thresholds when charging with mark-ups over short-run marginal costs

	Alternative 1: Air transport	*Alternative 2*: HSR	*Alternative 3*: Air transport and HSR
Access prices (€)	$\mu_a = 12$	$\mu_r = 50$	$\mu_a = 12; \mu_r = 50$
Ticket-prices (€)	$p_1 = p_2 = 70$	$p_r = 130$	$p_1 = p_2 = 65.30;$ $p_r = 93, 17$
Number of trips per user	$x_1 + x_2 = 100$	$x_r = 50$	$x_1 + x_2 = 83; x_r = 29$
First year social welfare functions	$\widehat{W}_a =$ 6500N-100,000,000	$\widehat{W}_r =$ 5250N-500,000,000	$\widehat{W}_{a+r} =$ 7686.8N-600,000,000
Minimum number of users	$N = 15{,}385$	–	$N = 421{,}300$
Minimum number of trips	$N(x_1 + x_2) =$ 1,538,500 Total: 1,538,500	–	$N(x_1 + x_2) =$ 34,854,149 $Nx_r = 12{,}182{,}311$ Total: 47,036,460

today and postpone the investment till the number of users is higher. However, if the number of users is between 15,385 and 421,300, the regulator should optimally choose *alternative 1*, i.e., to construct just the air infrastructure. If the number of users in the economy is higher than 421,300 the optimal choice is *alternative 3*, i.e., to construct both the air and rail infrastructures. *Alternative 2* (to construct just the rail infrastructure) is never optimal.

Notice that when moving from a charging scheme based on short-run marginal costs to a charging scheme based on mark-ups over short-run marginal costs, the first year social welfare function associated with each alternative becomes flatter. However, since $A < R$, the effect on the slope of the first year social welfare function is higher in the case of the rail infrastructure (see Fig. 4.4 in Sect. 4.3), and we need proportionally much more users for *alternative 3* to be chosen. Thus, charging according to short-run marginal costs favours the construction of the rail infrastructure.

When charging the use of transport infrastructures according to mark-ups over short-run marginal costs, the regulator aims to cover marginal operating and main-taining costs and at least part of the construction costs. In our example, the minimum number of users required for *alternative 3* to be optimal is 421,300. These users' demand 34,854,149 trips by air and 12,182,311 trips by HSR. With access charges, users cover all marginal operating and maintaining costs of the transport infrastruc-tures and part of the construction costs. In particular, they cover €70 million of the €100 million opportunity cost associated with the construction of the air trans-port infrastructure, and around €365 million of the €500 million opportunity cost associated with the construction of the HSR infrastructure.

4.5   Conclusions

Pricing and investment decisions must be taken together. They are not independent. When analysing different investment alternatives, we must compare the change in social surplus associated with them. Prices affect demand and, thus, surpluses and social welfare. Since social welfare is affected by prices, before deciding whether or not to invest in a project, we need to know the charging scheme. Moreover, different pricing policies may even change the investment decision due to the implied changes in social surplus. In this chapter, we show that certain charging schemes may favour the construction of certain transport infrastructures, leading to long-term equilibria that would not be optimal under other charging schemes.

To do so, in this chapter we use a theoretical model in which users' demand services in two transport modes: air transport and railway. Given users' preferences, the regulator must decide whether to invest in air transport, railway, both transport modes, or postpone the investment, using two possible charging schemes: either charging according to short-run marginal costs or charging mark-ups over short-run marginal costs in order to cover part of the construction costs. In this context, we show that the government may favour the construction of the rail infrastructure by choosing a charging scheme based on short-run marginal costs.

Sunk costs and irreversibility are higher for the rail infrastructure. On the one hand, the cost of building airports varies substantially with the level of demand, since the higher the level of demand, the higher the size of the airport. On the contrary, the cost of constructing the rail infrastructure varies little with the level of demand since the costliest part of such infrastructure is the rail tracks. On the other hand, once two regions have been connected with airports, only one more airport is needed to connect a third region (half of the previous investment). On the contrary, once two regions have been connected by rail, the cost of connecting the third region is almost the same. Moreover, airports allow regions to be connected by either short, medium, or long-haul flights, while HSR is only competitive for distances approximately below 800 km.

The practical consequence of this analysis for transport policy is straightforward: Investment planning at a government level should not be separated by product, like air transport and rail transport. It is common that public agencies at different government levels are organized by product instead of by function, with independent planning without due coordination. HSR infrastructure should be constructed only in those cases in which the level of social welfare is clearly higher than the social welfare associated with the next best alternative. This only happens when the level of the demand is sufficiently high, a fact that strongly depends on the charging scheme. The long-term consequences of investing in suboptimal infrastructure projects can be paramount. It may well be that this is not the optimal network but the irreversibility of the investment converts this suboptimal state in a long-term equilibrium. Once the infrastructure has been constructed, it should be used (if at least variable costs are lower than social benefits. This does not mean that new segments should be added to the existing network. The planner should wait till the demand reaches the required threshold for social profitability. Meanwhile, postponing the expansion of the network is socially worthy.

Appendix

In this chapter, we consider three alternative transport projects: to construct just the air transport infrastructure, to construct just the rail infrastructure, or to construct both the air and rail infrastructure. For the sake of simplicity, we assume infinite-life projects. Thus, the NPV_s of each alternative project can be written as:

$$NPV_s = -I + \frac{\overline{B}}{r_S}, \tag{4.22}$$

where I is the construction cost, $\overline{B}$ are annual net social benefits which are constant during the whole life of the project, and r_S is the social discount rate. In this case, $NPV_s > 0$ implies that $\overline{B} > r_S I$.

The decision about constructing today instead of postponing the investment (*optimal timing*) is taking considering the following expression:

$$\frac{\overline{B_1}}{1 + r_S} > \frac{r_S I}{1 + r_S}, \tag{4.23}$$

where $\overline{B_1}$ represents net social benefits during the first year. In other words, if net social benefits during the first year are higher than the opportunity cost of the investment, we should construct the infrastructure today. If annual net social benefits are constant, $\overline{B} = \overline{B_1}$, and we consider infinite-life projects, optimal timing condition is satisfied if and only if $NPV_s > 0$.

If annual net social benefits are not constant and the growth rate is $\theta < r_S$, $NPV_s > 0$ does not imply that the optimal timing condition is satisfied, since now the NPV_s is given by:

$$NPV_s = -I + \frac{\overline{B_1}}{r_S - \theta}. \tag{4.24}$$

In this case, it might be the case that the NPV_s is positive $(\overline{B_1} > (r_S - \theta)I)$ but the optimal timing condition is not satisfied since $\overline{B_1} < r_S I$. However, notice that if the optimal timing condition is satisfied $\overline{B_1} > r_S I$, then the NPV_s is positive since $\overline{B_1} > r_S I > (r_S - \theta)I$.

In the model used in this chapter, the first year social welfare is defined as first year net social benefits (obtained as the sum of consumers' surplus, transport operators' profits and the surplus due to the use of transport infrastructures) minus the investment opportunity cost. With this definition and considering infinite-life infrastructure projects, a positive first year social welfare implies, on the one hand, that it is optimal to construct the transport infrastructure today instead of postponing the investment one period and, on the other hand, that the NPV_s associated with such an infrastructure is positive.

Chapter 5
Cost–Benefit Analysis of Transport Policies: An Application to Subsidies in Air Transport Markets

5.1 Introduction

The cost-benefit analysis (CBA) of any public intervention requires the comparison of the situations with and without intervention, identifying winners and losers, and carefully analysing the key parameters that would allow the public policy to induce the desired effects in the market. Public intervention in transport markets may include not only the regulation of competition, safety standards, or the ownership of transport infrastructures, but also the subsidization of companies and/or passengers under the idea that transport services are essential, and citizens deserve a minimum supply of service at affordable prices.

In this chapter, we use the CBA in order to evaluate subsidies in transport markets. In particular, we focus on the effects of air transport subsidies in remote areas.[1] The common justification for this public assistance can be encapsulated in the improvement of transport connectivity and fair treatment of people living in islands and/ or remote locations. Although this is not necessarily the case, it is assumed that isolated areas and/or insular communities do not gather the minimum conditions for an adequate free-market supply of air transport services at affordable prices.

There are different policies that governments may implement in order to increase air connectivity in remote areas such as public service obligations (PSO), that can be open or restricted; subsidies to promote the start-up of new routes, usually granted to airlines through a competitive tendering process; subsidies to airports, that take the form of either discounts on airport charges paid by airlines that operate specific routes to remote areas, or direct or cross subsidies to airports located in remote regions; and subsidies for resident passengers. Subsidies for resident passengers may take the form of either an *ad valorem* (percentage of discount on the ticket-price) or a specific subsidy (fixed amount per trip irrespective of the price). Examples of this kind of subsidies can be found in European countries such as France, Greece, Italy, Portugal, Scotland, or Spain. While in Scotland and Spain, the subsidy for residents

---

[1] This chapter is based on de Rus & Socorro (2022).

© The Author(s), under exclusive license to Springer Nature Switzerland AG 2023
G. de Rus et al., *Economic Evaluation of Transport Projects*,
https://doi.org/10.1007/978-3-031-35959-0_5

is *ad valorem*, in Greece and Portugal the subsidy for residents is the difference between the market price and a flat rate. In France and Italy, residents also enjoy some benefits (such as discounts, maximum fares, flat rates) but, unlike in the other cases, all these benefits are included in PSO contracts.

In Sect. 5.2, we describe the main assumptions of the model and the main results in two extreme cases: the perfect competition and the monopoly. In Sect. 5.3, we describe the CBA methodology to analyse policies aimed to increase the air connectivity of resident passengers. In Sect. 5.4, we analyse the efficiency of *ad valorem* and specific subsidies for resident passengers. In order to do so, we should compare the change in social welfare. Since the aim of the policy is to favour resident passengers without harming non-residents, the public expenditure and surplus of all agents should be computed separately, paying special attention to the surplus of target groups. Therefore, we will follow the change in surpluses approach. In Sect. 5.5, we provide some empirical evidence and numerical simulations. Finally, Sect. 5.6 summarizes the main conclusions of this chapter.

5.2 The Basic Model

Let us consider N consumers willing to travel from region X to region Z, where region X is a remote region. Let $\alpha \in (0, 1]$ represent the proportion of passengers residing in region X (resident passengers). All resident passengers are assumed to be identical in their travel preferences, each of them with a linear downward-sloping inverse demand given by:[2]

$$p_d^R = a_R - b_R x^R, \tag{5.1}$$

where a_R and b_R are positive parameters representing the maximum willingness to pay (WTP) and the slope of the inverse demand function of a resident passenger, respectively. The term p_d^R represents the ticket-price paid by resident passengers, and x^R is the number of trips demanded by a representative resident passenger during a certain period of time. The higher is the value of b_R, the more price-inelastic is the demand function of the resident passenger.

Non-resident passengers are also assumed to be identical, each of them with a linear down-sloping inverse demand given by:

$$p_d^{NR} = a_{NR} - b_{NR} x^{NR}, \tag{5.2}$$

---

[2] For the sake of simplicity, we consider linear demand functions. However, main results also hold for non-linear demands, especially those related to the superiority of specific subsidies over *ad valorem* ones and how these differences may be mitigated when the subsidy is granted only to resident passengers.

where a_{NR} and b_{NR} are positive parameters representing the maximum WTP and the slope of the inverse demand function of a non-resident passenger, respectively. The term p_d^{NR} represents the ticket-price paid by non-resident passengers, and x^{NR} is the number of trips demanded by a representative non-resident passenger during a certain period of time. The higher is the value of b_{NR}, the more price-inelastic is the demand function of the non-resident passenger.

Notice that the maximum WTP of the representative resident passenger may be higher, lower or equal than the maximum WTP of the representative non-resident passenger. Similarly, the demand function of the representative resident passenger may be more or less price-inelastic than the demand function of the representative non-resident passenger. Although in the economic model we do not assume any specific value for these parameters, for consistency in all figures of this chapter, we represent a resident passenger with higher maximum WTP and more price-inelastic demand function than the one of the non-resident passenger. This is only due to exposition purposes and, as already highlighted, it does not correspond to any assumption or constraint on the values that can take these parameters.

In order to maintain the basic model as simple and intuitive as possible, we will consider only two extreme situations regarding the market structure: either a situation in which there are so many airlines operating in route XZ that none of them has any market power (the perfect competition case), or a situation in which just one airline operates in route XZ and, thus, such an airline has the maximum market power (the monopoly case). Any other real situation that might be considered regarding the market structure is between these two extreme cases. For the sake of simplicity, marginal operating costs in all cases are assumed to be constant and equal to c, with $c \geq 0$.[3]

5.2.1 Equilibrium in the Absence of Public Subsidies

The Perfect Competition Case

In the absence of any public intervention and perfect competition in the air transport market, the ticket-price paid by consumers coincides with the ticket-price charged by airlines, p_S^0, which in equilibrium is equal to the marginal operating cost, i.e., $p_d^R = p_d^{NR} = p_S^0 = c$. By substituting the equilibrium prices in expressions (5.1) and (5.2), we can obtain the quantities demanded by each resident and non-resident passenger in the equilibrium without subsidies, which are given by:

$$x_0^R = \frac{a_R - c}{b_R}, \quad x_0^{NR} = \frac{a_{NR} - c}{b_{NR}}. \tag{5.3}$$

---

[3] The assumption of constant marginal operating costs in air transport is quite common in the economic literature (Oum & Waters, 1997). However, if there is an expected increase of the demand, airlines might face increasing marginal operating costs in the short-run. We discuss this possibility later on.

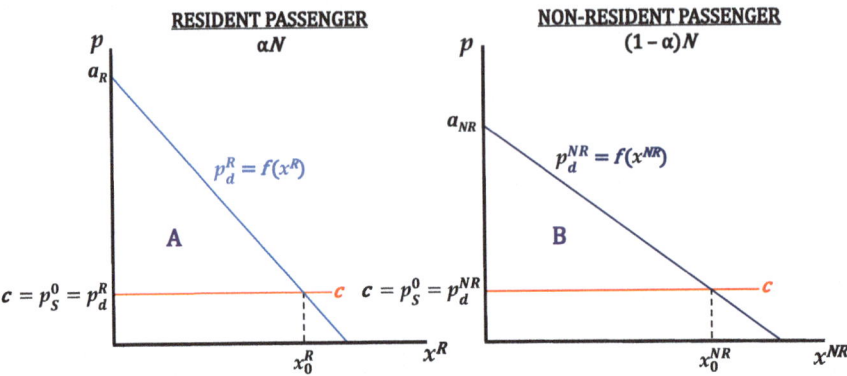

Fig. 5.1 Market equilibrium in the absence of public subsidies: the perfect competition case

Figure 5.1 illustrates the market equilibrium for the perfect competition case in the absence of any public intervention. When the ticket-price is set equal to the marginal operating cost, producers' surplus[4] is equal to zero, and resident consumers' surplus (CS^R) and non-resident consumers' surplus (CS^{NR}) are given by A and B, respectively, multiplied by the corresponding number of passengers:

$$CS^R = \alpha N A = \frac{1}{2}\alpha N (a_R - c) x_0^R,$$

$$CS^{NR} = (1 - \alpha) N B = \frac{1}{2}(1 - \alpha) N (a_{NR} - c) x_0^{NR}. \qquad (5.4)$$

The Monopoly Case

In the absence of any public intervention, the ticket-price paid by consumers coincides with the ticket-price charged by the airline. In this case, the monopoly carrier chooses the ticket-price p_S that maximizes his profits. Notice that this price is the same for residents and non-residents, since the monopolist is not allowed to price discriminate according to passengers' place of residence. In other words, the monopolist solves the following maximization program:

$$\max_{p_S} \alpha N (p_S - c) x^R + (1 - \alpha) N (p_S - c) x^{NR}, \qquad (5.5)$$

where x^R and x^{NR} are the number of trips demanded by residents and non-residents given by expressions (5.1) and (5.2), respectively, with $p_d^R = p_d^{NR} = p_S$.

---

[4] In this chapter, producers' surplus refers to both capital owners' surplus and landowners' surplus (see Chap. 2).

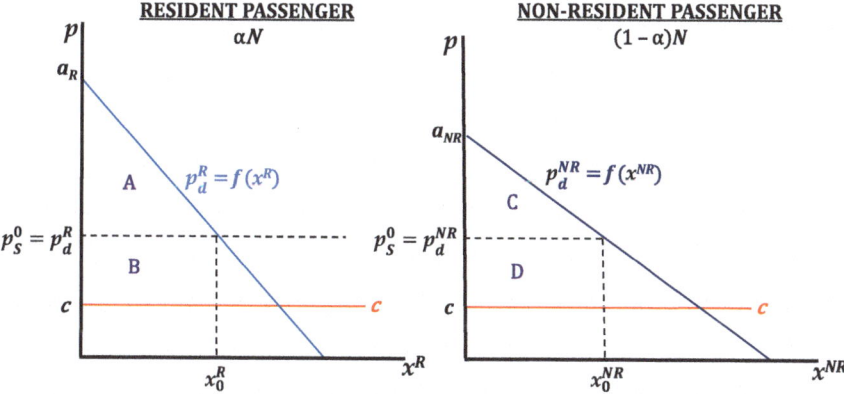

Fig. 5.2 Market equilibrium in the absence of public subsidies: the monopoly case

The first-order condition of the above maximization program yields the optimal ticket-price charged by the airline, p_S^0. By substituting such an optimal ticket-price p_S^0 in the demand functions given by expressions (5.1) and (5.2), and taking into account that in the absence of any public intervention $p_d^R = p_d^{NR} = p_S$, we can obtain the corresponding optimal quantities demanded by residents, x_0^R, and by non-residents, x_0^{NR} (mathematical expressions are given in Appendix 1).

Figure 5.2 illustrates the market equilibrium for the monopoly case in the absence of any public intervention. When the monopoly sets the ticket-price equal to p_S^0, resident consumers' surplus (CS^R), non-resident consumers' surplus (CS^{NR}), and producers' surplus are given by the following areas represented in Fig. 5.2, where each area is multiplied by the corresponding number of passengers:

$$CS^R = \alpha N A = \frac{1}{2}\alpha N \left(a_R - p_S^0\right)x_0^R,$$

$$CS^{NR} = (1 - \alpha)N C = \frac{1}{2}(1 - \alpha)N\left(a_{NR} - p_S^0\right)x_0^{NR},$$

$$PS = \alpha N B + (1 - \alpha)N D = \alpha N\left(p_S^0 - c\right)x_0^R + (1 - \alpha)N\left(p_S^0 - c\right)x_0^{NR}. \quad (5.6)$$

5.3 Cost–Benefit Analysis of Policies Aimed at Ensuring Air Connectivity of Resident Passengers

The CBA of any policy aiming at ensuring an adequate level of air connectivity and mobility of residents requires measuring the change in social welfare due to the policy. The change in social welfare can be defined as the weighted sum of the change in consumers' surplus (including resident and non-resident passengers), producers' surplus, taxpayers' surplus, and the surplus of the rest of society:[5]

$$\Delta W = \beta_R \Delta \text{CS}^R + \beta_{\text{NR}} \Delta \text{CS}^{\text{NR}} + \beta_P \Delta \text{PS} + \beta_G \Delta \text{GS} + \beta_E \Delta \text{ES}, \qquad (5.7)$$

where β_R, β_{NR}, β_P, β_G, and β_E represent the weight in social welfare of resident consumers' surplus, non-resident consumers' surplus, producers' surplus, taxpayers' surplus, and the surplus of the rest of the society, respectively. The change in the surplus of the rest of the society includes all possible (positive and/or negative) externalities that the policy may produce in the economy.

When income distribution is optimal, or the society has at its disposal means for unlimited and costless redistributions, the weights in the social welfare described in expression (5.7) can be set equal to one ($\beta_R = \beta_{\text{NR}} = \beta_P = \beta_G = \beta_E = 1$) and, thus, monetary gains and losses can be summed across individuals. However, redistribution is not necessarily costless since, for example, it might affect incentives in a negative way. In this case, the actual income distribution may not be far from the constrained optimal one. This means that the actual situation represents a kind of constrained optimum and possibly we can just sum gains and losses across individuals. This is also sufficient if relative prices are left more or less unchanged.[6]

Another possibility is simply to report the unweighted sum of gains and losses and allow the decision maker with the possibility to insert his own weights in the social welfare function. This is the approach we follow in this chapter. Taking into account that our purpose is to evaluate policies aimed at ensuring an adequate level of air connectivity and mobility of residents, as we will see later on, considering the weights in the social welfare function that would support such an objective would reinforce even more our conclusions.

Let us assume in this section that the policy only affects consumers' surplus, producers' surplus, and taxpayers' surplus. The change in all these surpluses due to the policy can be obtained by considering the prices and quantities before and after implementing the policy. In order to exemplify how to do so, let us analyse the situation illustrated in Fig. 5.3. The optimal ticket-price and quantities demanded by residents and non-residents in the absence of any public intervention are given by p_S^0, x_0^R and x_0^{NR}, respectively. Suppose now that the government introduces a discount for residents and the ticket-price increases till p_S^1. Residents pay the ticket-price minus the discount, i.e., $p_d^R < p_S^1$, while non-residents pay the whole ticket-price, i.e., $p_d^{\text{NR}} < p_S^1$. The government pays the difference between the ticket-price charged by the airline, p_S^1, and the price finally paid by residents, p_d^R.

We will now analyse how resident consumers' surplus, non-resident consumers' surplus, producers' surplus, and taxpayers' surplus change as a result of this policy.

---

[5] In this chapter, we consider no change in workers' surplus.

[6] See Johansson & Kriström (2016) for a detailed explanation of the aggregation problems that may arise and the practical approaches to address them.

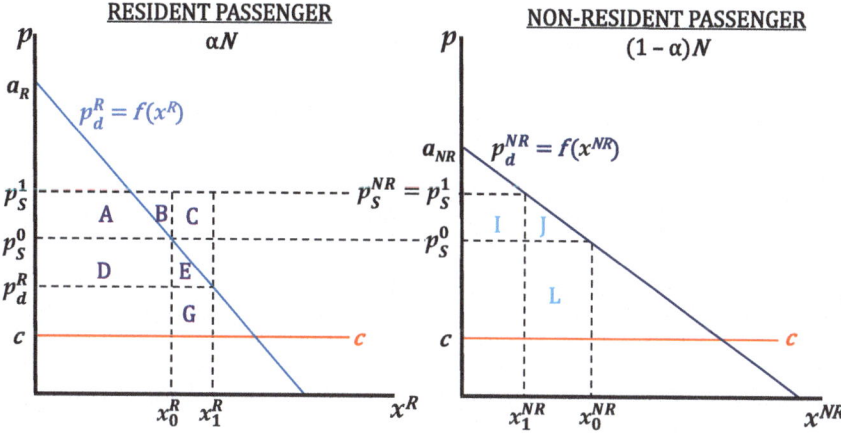

Fig. 5.3 Change in consumers' surplus, producers' surplus, and taxpayers' surplus due to the policy

The change in resident consumers' surplus, non-resident consumers' surplus, producers' surplus, and taxpayers' surplus due to this policy is computed taken into account the areas in Fig. 5.3, multiplied by the corresponding number of resident and non-resident passengers in the route.

Let us denote by TS the total surplus, which is defined as the sum of consumers' surplus, producers' surplus, and taxpayers' surplus. Notice that the change in total surplus coincides with the change in social welfare when all the weights in the social welfare function described in expression (5.7) are equal to one, i.e., $\beta_R = \beta_{NR} = \beta_P = \beta_G = 1$. The change in all surpluses due to this policy is then given by:

$$\Delta CS^R = \alpha N(D + E),$$
$$\Delta CS^{NR} = -(1 - \alpha)N(I + J),$$
$$\Delta PS = \alpha N(A + B + C + E + G) + (1 - \alpha)N(I - L),$$
$$\Delta GS = -\alpha N(A + B + C + D + E),$$
$$\Delta TS = \alpha N(E + G) - (1 - \alpha)N(J + L). \qquad (5.8)$$

The change in the total surplus can be also obtained as the difference between the change in WTP and the change in the use of the resources. This is shown in Fig. 5.4. The policy implies an increase in the number of trips from x_0^R to x_1^R for a resident passenger, and from x_0^{NR} to x_1^{NR} for a non-resident passenger, which implies an increase in the WTP equal to E, G, and M for a resident passenger and a decrease in the WTP equal to J, L, and Q for a non-resident passenger. From these areas, we have to consider the change in the use of resources; i.e., we have to subtract M (the cost of operating the new trips) for a resident passenger and add Q (saved cost) for a non-resident passenger. Finally, we have to multiply the areas by the corresponding number of passengers of each type, αN residents and $(1 - \alpha)N$ non-residents:

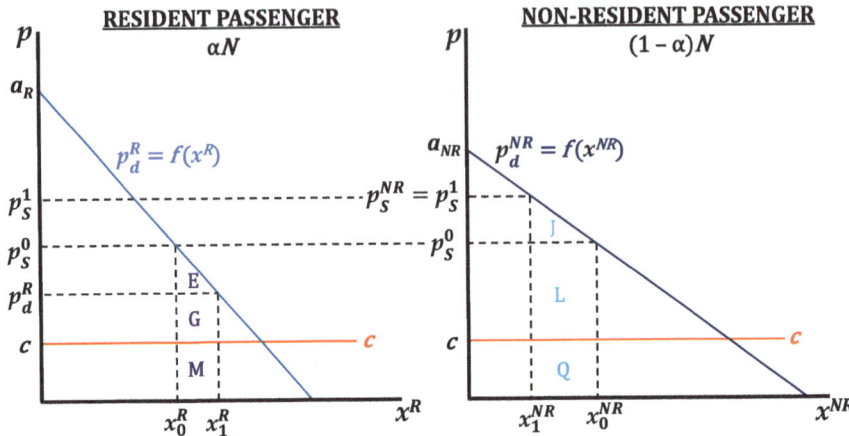

Fig. 5.4 Change in social surplus as the difference between the change in WTP and the use of resources

$$\Delta TS = \alpha N(E + G + M - M) - (1 - \alpha)N(J + L + Q - Q)$$
$$= \alpha N(E + G) - (1 - \alpha)N(J + L). \tag{5.9}$$

Although the total surplus given by expressions (5.8) and (5.9) coincide, in order to evaluate a policy aimed to residents, we have to clearly identify winners and losers and carefully analyse the key parameters that would allow the public policy to induce the desired effects in the market. For these purposes, the approximation given by the sum of surpluses seems more appropriate and, therefore, it is the one that we will use throughout the rest of this chapter.

Although, in order to evaluate the effects of transfers to residents, we will focus on the analysis of the change that such a policy implies on consumers' surplus, producers' surplus, and taxpayers' surplus in the direct market, there might be additional effects in other markets and/or externalities that might be considered in a sound CBA.

First, public funds are usually obtained through distortionary taxation and, thus, the economic cost of public funds should be included in the evaluation. The economic cost of public funds refers to the distortions created in the economy when raising taxes to finance the public policy.

Second, an increase in the number of passengers and/or flights due to the policy may produce more queues, delays, and/or congestion in airports. On the contrary, an increase in the number of flights due to the policy may produce a positive effect: the impact on the so-called schedule delay. Passengers have a preferred departure time and dislike the schedule delay, which is the difference between the actual and preferred departure time. An increase in the numbers of flights (frequency) reduces the schedule delay and, hence, the consumers' generalized price.

Third, an increase in the number of flights may produce a negative environmental impact in terms of noise and air pollution. The negative environmental impact due to an increase in air traffic may depend on a set of variables such as the type of aircraft used by airlines and the existence of population living near the airport, and thus, it should be analysed case by case.

Fourth, the policy may affect airlines' operating costs, either due to the need to attend the increased demand in the short-run, or because the policy affects airlines' incentives to be cost efficient. On the one hand, marginal operating costs may be increasing in the short-run. Since the policy may imply an increase in the number of flights to be operated by airlines, the airlines may face an increase in their operating costs to attend such an increased demand in the short-run (more and more costly aircraft, more and more costly crew, etc.). On the other hand, if due to the policy, airlines lose their incentives to be cost efficient and increase their operating costs, such a negative effect should be also considered in the CBA.

Finally, the policy may have a positive impact on the level of competition. The policy may have a positive effect in incumbents' profits that may attract new entrants to the market. If that is the case, the CBA should consider the new market structure when evaluating the situation with the policy in comparison with the situation without the policy. Moreover, not only intramodal competition should be considered but also intermodal competition. When talking about competition in the route we refer not only to the number of airlines operating the route but also other transport modes that passengers may use to move from region X to region Z (e.g., maritime transport).

5.4 Comparison of Different Policies Aimed at Ensuring Air Connectivity of Resident Passengers

5.4.1 Ad Valorem *Subsidies Only for Residents*

An *ad valorem* subsidy for resident passengers is a subsidy based on the ticket-price paid by passengers living in region X. Let us denote by $\sigma \in (0, 1)$ the proportion of the ticket-price that is subsidized. Since resident passengers are entitled to receive the subsidy, the ticket-price finally paid by those passengers is equal to the ticket-price charged by the carrier minus the *ad valorem* subsidy, i.e., $p_d^R = (1 - \sigma)p_S$. On the contrary, the ticket-price paid by non-residents is just the ticket-price charged by the carrier, i.e., $p_d^{NR} = p_S$.

The Perfect Competition Case

With perfect competition, the ticket-price charged by airlines in equilibrium is equal to the marginal operating cost, i.e., $p_S^1 = p_S^0 = c$. Thus, the ticket-price charged by airlines is exactly the same that before the subsidy. Since residents pay a ticket-price equal to the ticket-price charged by airlines minus the discount, $p_d^R = (1 - \sigma)p_S$, they finally pay a lower price than before the subsidy, $p_d^R = (1 - \sigma)c$, and demanding more. On the contrary, non-residents end up paying exactly the same ticket-price that before the subsidy, $p_d^{NR} = p_S^1 = p_S^0 = c$, and demanding the same amount:

$$x_1^R = \frac{a_R - c(1 - \sigma)}{b_R}, \qquad x_1^{NR} = x_0^{NR} = \frac{a_{NR} - c}{b_{NR}}. \tag{5.10}$$

In perfect competition, the ticket-price charged by airlines is set equal to the marginal operating cost and, thus, producers' surplus is zero. Since the ticket-price charged by airlines is unaltered and non-residents receive no subsidy, the change in non-resident consumers' surplus is zero. The change in resident consumers' surplus, and taxpayers' surplus due to this policy are given by the areas in Fig. 5.5, multiplied by the corresponding number of resident and non-resident passengers in the route. With perfect competition, the change in total surplus is negative and equal to C, multiplied by the number of residents:

$$\Delta CS^R = \alpha N(A + B),$$
$$\Delta GS = -\alpha N(A + B + C),$$
$$\Delta TS = -\alpha NC. \tag{5.11}$$

The subsidy for residents implies an efficiency loss equal to αNC representing that for the new trips, $\alpha N\left(x_1^R - x_0^R\right)$, residents are willing to pay less than the cost of operating those additional trips.

It is worth highlighting that, though A and B are transfers from taxpayers to each resident passenger and, thus, they are cancelled in the computation of the change in total surplus, they might be multiplied by different weights in the social welfare

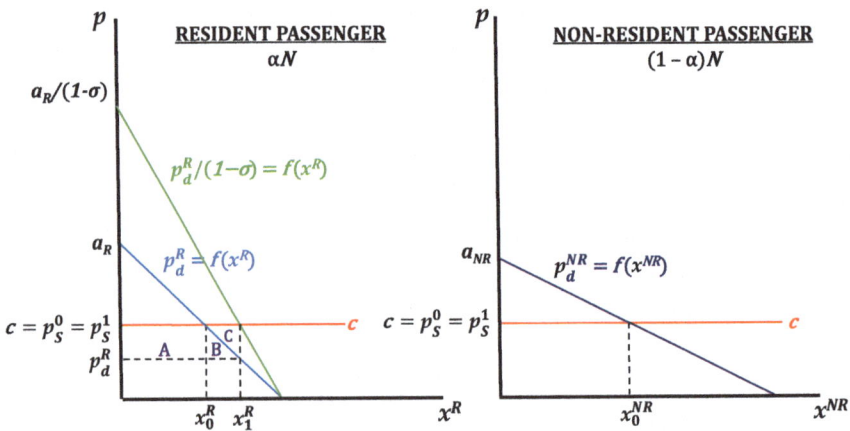

Fig. 5.5 Change in consumers' surplus, producers' surplus, and taxpayers' surplus due to an *ad valorem* subsidy for residents: the perfect competition case

function. This means that from the social point of view A and B might not be treated as mere transfers, and if $\beta_R > \beta_G$ the change in social welfare might be positive, even though the change in total surplus is undoubtedly negative.

The Monopoly Case

When the route is operated by a monopolist and an *ad valorem* subsidy only for residents is introduced, the airline solves the following maximization program:

$$\max_{p_S} \alpha N(p_S - c)x^R + (1 - \alpha)N(p_S - c)x^{NR}, \tag{5.12}$$

where x^R and x^{NR} represents, given the price that they finally pay, the quantity demanded by residents and non-residents, respectively. Thus, $x^R = \frac{a_R - (1-\sigma)p_S}{b_R}$ and $x^{NR} = \frac{a_{NR} - p_S}{b_{NR}}$.

The first-order condition of the above maximization program yields the optimal ticket-price charged by the airline, p_S^1. By substituting such an optimal ticket-price p_S^1 in the corresponding demand functions, we can obtain optimal quantities demanded by residents, x_1^R, and by non-residents, x_1^{NR} (see all mathematical expressions in Appendix 1).

Figure 5.6 summarizes the effects of an *ad valorem* subsidy for residents for the monopoly case. An *ad valorem* subsidy affects the slope of the residents' demand function. This fact is taken into account by the monopolist that faces the trade-off between increasing the ticket-price to appropriate as much as possible of the residents' subsidy while losing non-residents demand, or maintaining the ticket-price close to the case in which there is no public intervention. The degree in which the ticket-price charged by the airline is increased, $p_S^1 - p_S^0$, will strongly depend on the proportion of resident passengers in the route, α, the amount of the *ad valorem* subsidy, σ, the maximum WTP of residents and non-residents, a_R. and a_{NR}, and the slope of residents' and non-residents' inverse demand functions, b_R and b_{NR} (mathematical expressions can be found in Appendix 1). The higher the maximum WTP of non-residents a_{NR} is, the higher the difference between $p_S^1 - p_S^0$ is. Similarly, the higher the proportion of resident passengers in the route α is, the higher the difference between $p_S^1 - p_S^0$ is. Thus, the size and the proportion of non-residents in the route are elements that undoubtedly mitigate the non-desirable effects in the market of the *ad valorem* subsidy for residents.

The change in resident consumers' surplus, non-resident consumers' surplus, producers' surplus, and taxpayers' surplus due to this policy are given by the areas in Fig. 5.6, multiplied by the corresponding number of resident and non-resident passengers in the route:

$$\Delta CS^R = \alpha N(D + E),$$
$$\Delta CS^{NR} = -(1 - \alpha)N(I + J),$$

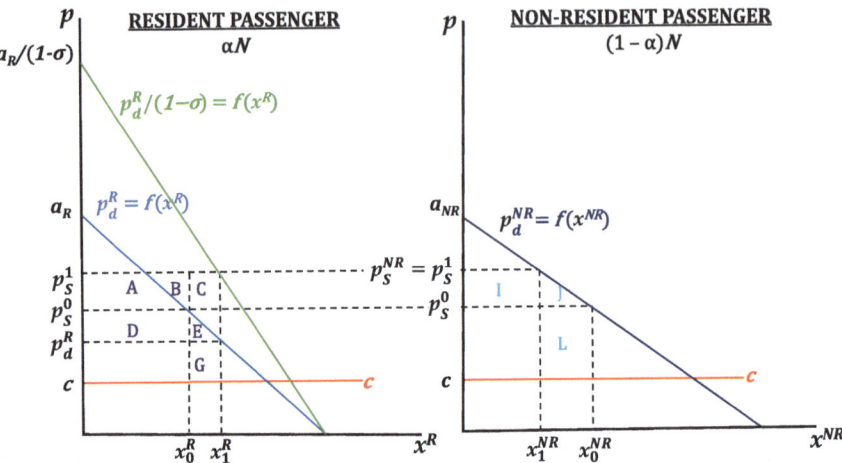

Fig. 5.6 Change in consumers' surplus, producers' surplus, and taxpayers' surplus due to an *ad valorem* subsidy for residents: the monopoly case

$$\Delta PS = \alpha N(A + B + C + E + G) + (1 - \alpha)N(I - L),$$
$$\Delta GS = -\alpha N(A + B + C + D + E),$$
$$\Delta TS = \alpha N(E + G) - (1 - \alpha)N(J + L). \qquad (5.13)$$

The change in total surplus reflects the efficiency of the *ad valorem* subsidy only for residents, meaning that the policy is efficient if the increase in the WTP minus the resources used for the increase in the number of trips of residents compensates the decrease in the WTP plus the saving in resources caused by the decrease in the number of trips of non-residents.

Again, it is worth highlighting that, though A, B, C, D, E, and I are transfers from one agent to the other and, thus, they are cancelled in the computation of the change in total surplus, they might be multiplied by different weights in the social welfare function. This means that the value and sign of the change in social welfare might be different than the value and sign of the change in total surplus.

5.4.2 Specific Subsidies Only for Residents

A specific subsidy only for residents consists of granting a fixed amount per ticket s to all passengers living in region X, independently on the ticket-price. Obviously, this fixed amount should be lower than the maximum ticket-price and, hence, we assume that $0 < s < a_R$. Since resident passengers are entitled to receive the subsidy, the ticket-price finally paid by those passengers is equal to the ticket-price charged by the carrier minus the specific subsidy, i.e., $p_d^R = p_S - s$.

The Perfect Competition Case

With perfect competition, the ticket-price charged by airlines in equilibrium is equal to the marginal operating cost, i.e., $p_S^1 = p_S^0 = c$. Thus, the ticket-price charged by airlines is exactly the same that before the subsidy. Since residents pay a ticket-price equal to the ticket-price charged by airlines minus the discount, $p_d^R = p_S - s$, they end paying a lower price than before the subsidy, $p_d^R = c - s$, and demanding more. On the contrary, non-residents end up paying exactly the same ticket-price that before the subsidy and demanding the same amount:

$$x_1^R = \frac{a_R - c + s}{b_R}, \quad x_1^{NR} = x_0^{NR} = \frac{a_{NR} - c}{b_{NR}}. \tag{5.14}$$

With perfect competition, the ticket-price charged by airlines is set equal to the marginal operating cost and, thus, producers' surplus is zero. Since the ticket-price charged by airlines is unaltered and non-residents receive no subsidy, the change in non-resident consumers' surplus is zero. The change in resident consumers' surplus and taxpayers' surplus due to this policy are given by the areas in Fig. 5.7, multiplied by the corresponding number of resident and non-resident passengers in the route:

$$\Delta CS^R = \alpha N(A + B),$$
$$\Delta GS = -\alpha N(A + B + C),$$
$$\Delta TS = -\alpha N C. \tag{5.15}$$

The subsidy for residents implies a efficiency loss equal to $\alpha N C$ representing that for the new trips, $\alpha N\left(x_1^R - x_0^R\right)$, residents are willing to pay less than the cost of operating those additional trips.

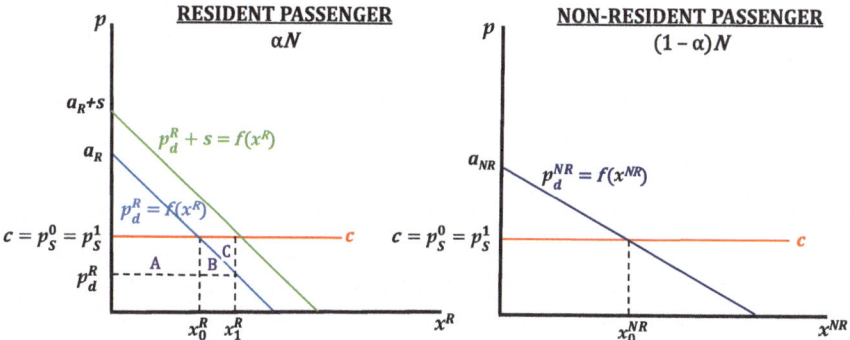

Fig. 5.7 Change in consumers' surplus, producers' surplus, and taxpayers' surplus due to a specific subsidy for residents: the perfect competition case

Similar to the *ad valorem* case, though A and B are transfers from taxpayers to each resident passenger and, thus, they are cancelled in the computation of the change in total surplus, they might be multiplied by different weights in the social welfare function. This means that from the social point of view A and B might not be treated as mere transfers.

The Monopoly Case

When the route is operated by a monopolist and a specific subsidy only for residents is introduced, the airline solves the following maximization program:

$$\max_{p_S} \alpha N (p_S - c) x^R + (1 - \alpha) N (p_S - c) x^{NR}, \qquad (5.16)$$

where x^R and x^{NR} represents, given the price that they finally pay, the quantity demanded by residents and non-residents, respectively. Thus, $x^R = \frac{a_R - p_S + s}{b_R}$. and $x^{NR} = \frac{a_{NR} - p_S}{b_{NR}}$.

The first-order condition of the above maximization program yields the optimal ticket-price charged by the airline, p_S^1. By substituting such an optimal ticket-price p_S^1 in the corresponding demand functions, we can obtain optimal quantities demanded by residents, x_1^R, and by non-residents, x_1^{NR} (see all mathematical expressions in Appendix 1).

The change in resident consumers' surplus, non-resident consumers' surplus, producers' surplus, and taxpayers' surplus due to this policy the areas in Fig. 5.8, multiplied by the corresponding number of resident and non-resident passengers in the route:

$$\Delta CS^R = \alpha N (D + E),$$
$$\Delta CS^{NR} = -(1 - \alpha) N (I + J),$$
$$\Delta PS = \alpha N (A + B + C + E + G) + (1 - \alpha) N (I - L),$$
$$\Delta GS = -\alpha N (A + B + C + D + E),$$
$$\Delta TS = \alpha N (E + G) - (1 - \alpha) N (J + L). \qquad (5.17)$$

The interpretation of these results is similar to the one of the *ad valorem* subsidy only for residents.

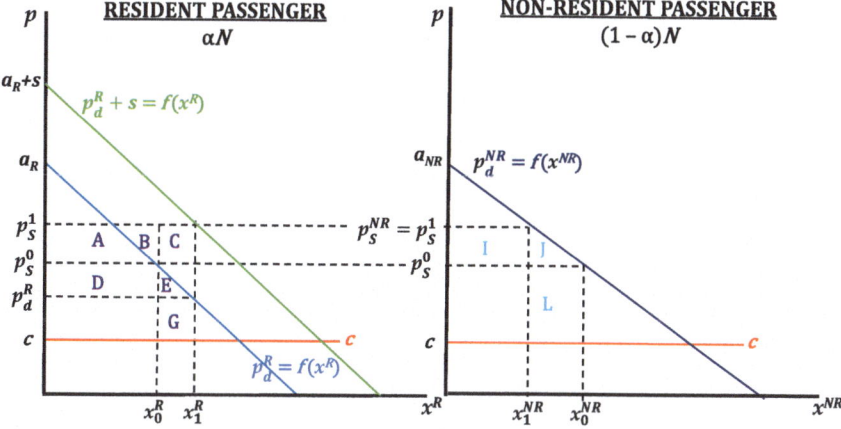

Fig. 5.8 Change in consumers' surplus, producers' surplus, and taxpayers' surplus due to a specific subsidy for residents: the monopoly case

5.4.3 Ad Valorem *Versus Specific Subsidies Only for Residents: The Effects on Prices and All Agents' Surpluses*

The Perfect Competition Case

With perfect competition in the air transport market, the ticket-price charged by airlines in equilibrium is equal to the marginal operating cost. Thus, when the government introduces a subsidy only for residents, the ticket-price charged by airlines is the same as before the subsidy. Since residents pay a ticket-price equal to the ticket-price charged by airlines minus the subsidy, they fully enjoy the subsidy and demand more. On the contrary, non-resident passengers end up paying the same ticket-price as before the subsidy. Therefore, with perfect competition in the air transport market, both the *ad valorem* and specific subsidy only for residents are fully effective: residents fully enjoy the subsidy while non-resident passengers end up paying the same ticket-price as before the subsidy:

$$x_1^{NR} = x_0^{NR} = \frac{a_{NR} - c}{b_{NR}}. \tag{5.18}$$

However, with an *ad valorem* subsidy residents pay a price equal to $c(1 - \sigma)$ while with a specific subsidy the price becomes $c - s$. Therefore, by setting a specific subsidy equal to $s = c\sigma$ the price finally paid by residents is exactly the same that the one paid by residents and an *ad valorem* subsidy, where the quantity demanded by a representative resident passenger with any of these subsidies is given by:

$$x_1^R = \frac{a_R - c + c\sigma}{b_R}.$$ (5.19)

With such a specific subsidy, resident consumers' surplus, non-resident consumers' surplus, producers' surplus, and the government's expenditure are identical than the corresponding surpluses obtained with an *ad valorem* subsidy, which are given by:

$$CS^R = \frac{1}{2}\alpha N(a_R - c + c\sigma)x_1^R,$$

$$CS^{NR} = \frac{1}{2}(1 - \alpha)N(a_{NR} - c)x_1^{NR},$$

$$PS = 0,$$

$$GS = -\alpha N c\sigma x_1^R,$$ (5.20)

where x_1^R and x_1^{NR} are given by expressions (5.18) and (5.19), respectively.

Figure 5.9 represents the specific subsidy granted only for residents that yields exactly the same results that the corresponding *ad valorem* one when the market is under perfect competition: $s = c\sigma$.

The Monopoly Case

Let us denote by $\left(p_S^1\right)^{AV}$ the ticket-price charged by the airline when an *ad valorem* subsidy only for residents is introduced, i.e., the price that solves the maximization program given by expression (5.12). Let $\left(p_S^1\right)^s$ represent the ticket-price charged by

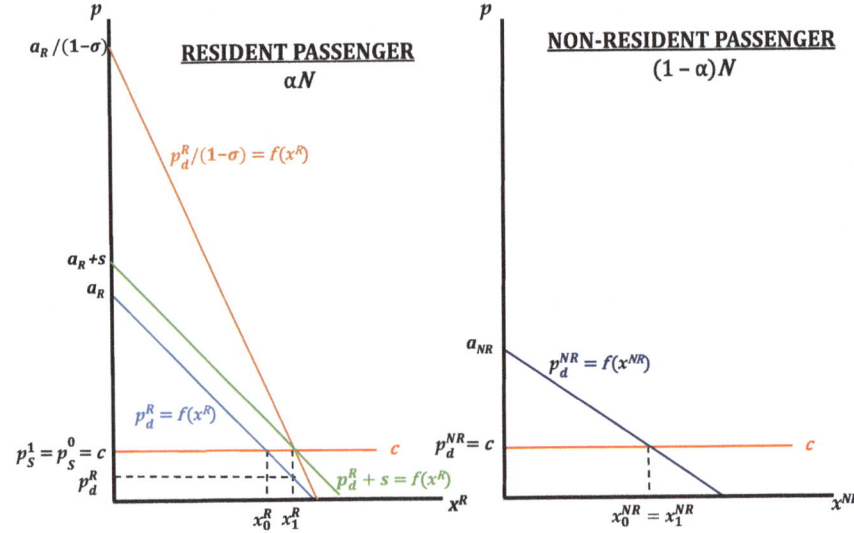

Fig. 5.9 Comparison of *ad valorem* and specific subsidy for residents: the perfect competition case

the airline when a specific subsidy only for residents is introduced, i.e., the price that solves the maximization program given by expression (5.16) (all mathematical expressions are in Appendix 1).

Let us compare the efficiency of the *ad valorem* and the specific subsidy for residents. Let us start comparing the efficiency of the *ad valorem* subsidy for residents and the specific subsidy that would allow resident passengers to pay the same price that they would pay with the *ad valorem* one. The specific subsidy for residents s_1^* that allows resident passengers to pay the same price that they would pay with an *ad valorem* subsidy of value σ is given by solving:

$$\left(p_d^R\right)^{AV} = \left(p_S^1\right)^{AV}(1-\sigma) = \left(p_S^1\right)^s - s_1^* = \left(p_d^R\right)^s, \tag{5.21}$$

whose solution is given by:

$$s_1^* = \frac{\sigma\left(b_R^2(1-\alpha)^2(c+a_{NR}) + \alpha^2 b_{NR}^2 c(1-\sigma) + \alpha(1-\alpha)b_{NR}b_R(2c+a_R-c\sigma)\right)}{(2b_R(1-\alpha)+\alpha b_{NR})(b_R(1-\alpha)+\alpha b_{NR}(1-\sigma))}. \tag{5.22}$$

As shown in Fig. 5.10, for such a specific subsidy s_1^* the ticket-price charged by the airline with an *ad valorem* subsidy is higher than the ticket-price charged with the specific subsidy, $\left(p_S^1\right)^{AV} > \left(p_S^1\right)^s$. Since non-residents are not entitled to receive the subsidy and $\left(p_S^1\right)^{AV} > \left(p_S^1\right)^s$, the quantity demanded by non-residents is higher with the specific subsidy: $\left(x_1^{NR}\right)^s > \left(x_1^{NR}\right)^{AV}$.

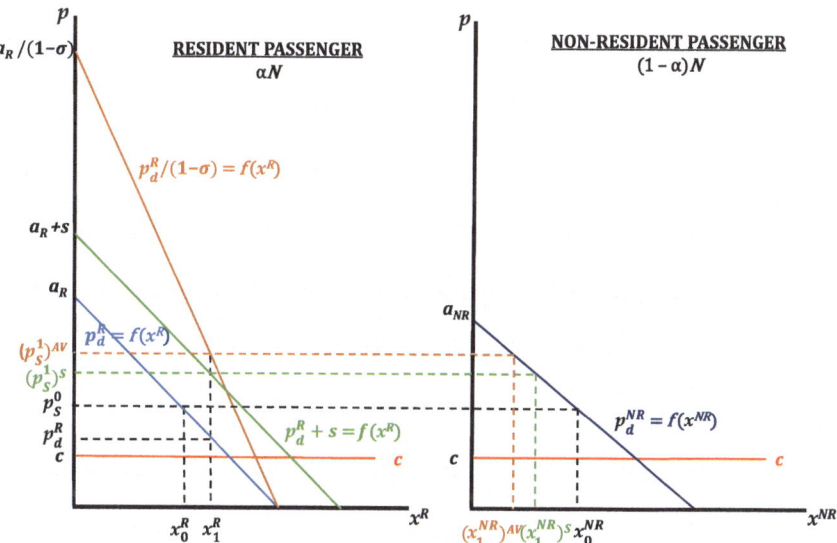

Fig. 5.10 Comparison of *ad valorem* and specific subsidy for residents: The monopoly case

As a consequence, with a specific subsidy s_1^* residents are equal, non-residents are better-off, and airline's profits and the government's expenditure are lower than with the *ad valorem* subsidy.

Alternatively, we may compare the efficiency of the *ad valorem* subsidy for residents and the specific subsidy that would imply the same government's expenditure. The specific subsidy for residents s_2^* that implies the same government's expenditure (GS) that an *ad valorem* subsidy of value σ is given by solving:

$$(\text{GS})^{\text{AV}} = -\alpha N \sigma \left(p_S^1\right)^{\text{AV}} \left(x_1^R\right)^{\text{AV}} = -\alpha N s_2^* \left(x_1^R\right)^s = (\text{GS})^s.$$

With a specific subsidy for residents s_2^* the government's expenditure is the same as before, residents and non-residents are better-off, and the airline's profits are lower than with the *ad valorem* subsidy.

When airlines have market power, the *ad valorem* subsidy is equivalent to the combined use of a specific subsidy plus a profit subsidy. Hence, although a specific subsidy leads to the same price for residents (or the same public expenditure) as the *ad valorem* one, the *ad valorem* subsidy increases the monopolist's market power and, thus, implies that the airline charges a higher price, resulting in lower non-resident consumers' surplus and higher profits for the airline. Therefore, when airlines have market power, the specific subsidy for residents is always superior in terms of social welfare to the *ad valorem* one.

5.5   Some Numerical Illustrations and Empirical Evidence

In order to illustrate the main insights of the model, we consider some numerical simulations. The results are compared with the empirical evidence provided by a study conducted by the Spanish *Independent Authority of Fiscal Responsibility* (AIReF, 2020). The study focuses on the analysis of the effects of the *ad valorem* subsidy applied to residents in the Canary and Balearic Islands and, in particular, on the effects of the increase from 50% to 75% discount on the ticket-price of all domestic and interisland flights bought by residents, increase that produced a growth in the public expenditure from €324 million in 2016 to €730 million in 2019. AIReF (2020) uses two databases, one with 2 million flights from July 2009 to June 2019 and another one with more than 100 million subsidized tickets from July 2009 to June 2019. Interislands flights in the Canary Islands are operated by a monopoly airline, while there is some competition in domestic non-interislands flights. In the Balearic Islands, there is some competition in both interislands and domestic non-interislands flights. Appendix 2 provides information on main subsidized routes from/to the Canary and Balearic Islands during the period July 2018–June 2019 (see Tables 5.6, 5.7, 5.8 and 5.9).

The economic model developed in this chapter provides theoretical support to the empirical findings obtained by AIReF (2020). Moreover, it allows us to simulate the effects of the *ad valorem* subsidy for residents and its equivalent specific subsidy, providing a measure of the economic consequences of choosing the wrong mechanism. To do so, let us start assuming the following values for the parameters of the model:

Numerical Example 1: $a_R = 100$, $b_R = 1$, $a_{NR} = 165$, $b_{NR} = 0.8$, $c = 10$, $N = 10,000$.

In this first numerical example, we are considering the case in which non-residents have an inverse demand function with similar slope than residents, but non-resident's maximum WTP is higher.

Firstly, let us analyse the effects on the ticket-price charged by the airline (and, therefore, paid by non-residents) when the government uses an *ad valorem* subsidy. Let us also simulate the effects of an increase in the *ad valorem* subsidy for residents from 50% to 75%. Table 5.1 summarizes the main results of such an increase in numerical example 1.

As shown in Table 5.1, the increase in the ticket-price charged by the airline due to an increase in the *ad valorem* subsidy crucially depends on the proportion of residents on the route, α. In particular, such an increase is higher, the higher is α. This result coincides with the empirical evidence provided by AIReF (2020), that analyses the effects of the increase in the *ad valorem* subsidy from 50% to 75% in the Canary and Balearic Islands. AIReF (2020) uses different econometric models (such as a regression discontinuity design or difference in differences models) that include the proportion of resident passengers as a key variable to be taken into account.

On the one hand, Table 5.2 shows the effects on the ticket-price charged by airlines due to an increase in the *ad valorem* subsidy for residents in domestic non-interislands flights from/to the Canary Islands during the period 2017–2019. Notice that the proportion of resident passengers in domestic non-interislands flights from/to the Canary Islands ranges from 14% to 78% on those routes (see Table 5.8 in Appendix 2). In order to perform the analysis, AIReF (2020) divides the number of passenger-trips in different quintiles according to the proportion of resident passengers on each route.

Table 5.1 Effects of an increase in the *ad valorem* subsidy for residents (numerical example 1)

	$\sigma = 0.5$			$\sigma = 0.75$		
	$\alpha = 0.1$	$\alpha = 0.5$	$\alpha = 0.9$	$\alpha = 0.1$	$\alpha = 0.5$	$\alpha = 0.9$
Ticket-price charged by the airline	88.24	92.5	101.20	90.05	107.08	163.04
Increase in the ticket-price charged by the airline (%)	–	–	–	2.05	15.76	61.10

Note: σ denotes the *ad valorem* subsidy and α the percentage of resident passengers

Table 5.2 Effects of an increase in the *ad valorem* subsidy for residents in domestic non-interislands flights from/to the Canary Islands during the period 2017–2019

	According to the percentage of resident passengers				
	Quintile 1	Quintile 2	Quintile 3	Quintile 4	Quintile 5
Increase in the ticket-price charged by the airline (%)	− 0.2	7.7	12	15.5	22.2

Source: AIReF (2020)

On the other hand, Table 5.3 analyses the effects on the ticket-price charged by airlines of an increase in the *ad valorem* subsidy for residents in domestic non-interislands flights from/to the Balearic Islands during the period 2017–2019, where the proportion of resident passengers ranges from 15% to 61% (see Table 5.9 in Appendix 2). Again, in order to perform the analysis, AIReF (2020) divides the number of passenger-trips in different quintiles according to the proportion of resident passengers on each route.

Let us now compare the efficiency of the *ad valorem* and the equivalent specific subsidy. Table 5.4 represents the results obtained in numerical example 1 in the absence of public intervention and when the subsidy (either *ad valorem* or specific) is given only to residents, considering different values of the proportion of resident passengers on the route, α. We analyse the case in which the government decides to introduce an *ad valorem* subsidy for residents equal to 75% discount on the ticket-price, $\sigma = 0.75$. The equivalent specific subsidy is computed in order to imply either the same price paid by residents that the *ad valorem* one—specific subsidy (1)—or, alternatively, the same public expenditure: specific subsidy (2).

We consider three possible values of α: 0.1, 0.5, and 0.9. The first two values usually correspond to flights in domestic non-interislands routes (see routes in Tables 5.8 and 5.9 in Appendix 2). The latter value, $\alpha = 0.9$, usually corresponds to flights in interislands routes (see routes in Tables 5.6 and 5.7 in Appendix 2).

Let us start by comparing the results of the *ad valorem* subsidy for residents and the specific subsidy (1). Notice that for every value of α, with the specific subsidy (1) residents are indifferent, but non-residents and taxpayers are better-off than with the *ad valorem* subsidy. On the contrary, the airlines' profits are lower. However, the magnitude of this superiority strongly depends on the value of α. To achieve

Table 5.3 Effects of an increase in the *ad valorem* subsidy for residents in domestic non-interislands flights from/to the Balearic Islands during the period 2017–2019

	According to the percentage of resident passengers				
	Quintile 1	Quintile 2	Quintile 3	Quintile 4	Quintile 5
Increase in the ticket-price charged by the airline (%)	5.4	10.3	13.2	16.0	30.7

Source: AIReF (2020)

Table 5.4 Effects of the *ad valorem* and specific subsidy only for residents for different values of α and $\sigma = 0.75$ (numerical example 1)

	$\alpha = 0.1$				$\alpha = 0.5$				$\alpha = 0.9$			
	Without subsidies	Ad valorem	Specific (1)	Specific (2)	Without subsidies	Ad valorem	Specific (1)	Specific (2)	Without subsidies	Ad valorem	Specific (1)	Specific (2)
p_S	84.85	90.05	87.50	87.56	73.06	107.08	86.28	88.91	58.96	163.04	73.21	95.31
p_d^R	84.85	22.51	22.51	21.17	73.06	26.77	26.77	17.56	58.96	40.76	40.76	12.51
ΔCS^R	–	2.89	2.89	3.01	–	11.59	11.59	15.17	–	8.21	8.21	28.86
ΔCS^{NR}	–	– 4.55	– 2.35	– 2.41	–	– 15.94	– 7.06	– 8.33	–	– 7.028	– 1.77	– 4.00
ΔPS	–	5.08	4.95	5.06	–	25.96	20.73	25.32	–	57.4	16.38	50.04
ΔGS	–	– 5.23	– 5.03	– 5.23	–	– 29.41	– 21.79	– 29.41	–	– 65.19	– 17.30	– 65.19

Note: σ denotes the *ad valorem* subsidy and α the percentage of resident passengers. p_S is the price charged by the airline and p_d^R the price paid by resident passengers. ΔCS^R is the change in resident consumers' surplus, ΔCS^{NR} the change in non-resident consumers' surplus, ΔPS the change in the airline's profits, and ΔGS the change in taxpayers' surplus

Specific (1): The specific subsidy that yields the same price for residents that the *ad valorem* one

Specific (2): The specific subsidy that yields the same government's expenditure that the *ad valorem* one

Prices are in euros, while changes in consumers' surplus, the airline's profits, and government's expenditure are measured in millions of euros

the same resident consumers' surplus (CS^R) that the equivalent specific subsidy—specific subsidy (1)—, with the *ad valorem* subsidy the government needs to expend around 4% more public resources if $\alpha = 0.1$, 34% more public resources if $\alpha = 0.5$, and around 277% more public resources if $\alpha = 0.9$. Besides, with the *ad valorem* subsidy, the increase in the airline's profits (PS) is 2.62% higher if $\alpha = 0.1$, 25.23% higher if $\alpha = 0.5$, and 250.42% higher if $\alpha = 0.9$. On the contrary, with the *ad valorem* subsidy the decrease in non-resident consumers' surplus (CS^{NR}) is 93.61% higher if $\alpha = 0.1$, 125.78% higher if $\alpha = 0.5$, and 297.06% higher if $\alpha = 0.9$. Therefore, the economic consequences of choosing the wrong mechanism are quite different depending on the proportion of resident passengers on the route.

Let us now compare the results of the *ad valorem* subsidy for residents and the specific subsidy (2). Notice that for every value of α, with the specific subsidy (2) the public expenditure is the same, but residents and non-residents are better-off than with the *ad valorem* subsidy. On the contrary, the airlines' profits are lower. However, the magnitude of this superiority strongly depends on the proportion of resident passengers. With the same public expenditure that the *ad valorem* subsidy for residents, the increase in resident consumers' surplus with the equivalent specific subsidy—specific subsidy (2)—is 4.15% higher if $\alpha = 0.1$, 30.89% higher if $\alpha = 0.5$, and 251.52% higher if $\alpha = 0.9$. As far as the airline's profits is concerned, compared to the *ad valorem* subsidy, with the equivalent specific subsidy the increase in profits is 0.4% lower if $\alpha = 0.1$, 2.46% lower if $\alpha = 0.5$, and 12.82% lower if $\alpha = 0.9$. Therefore, once again, we can conclude that the economic consequences of choosing the wrong mechanism are quite different depending on the proportion of resident passengers on the route.

Let us now consider the following values for the parameters:

Numerical Example 2: $a_R = 100$, $b_R = 1$, $a_{NR} = 50$, $b_{NR} = 0.2$, $c = 10$, $N = 10,000$.

In this second example, we are considering the case in which non-residents have a very flat inverse demand function and their WTP is low in comparison with residents'. Let us compare the results with numerical example 1, i.e., the case in which non-residents have an inverse demand function with similar slope than residents' (thought non-resident's maximum WTP is higher). In order to make the comparisons, we consider that the proportion of resident passengers is $\alpha = 0.5$ and the government decides to introduce an *ad valorem* subsidy for residents equal to 75% discount on the ticket-price, $\sigma = 0.75$.

Table 5.5 summarises the results obtained in numerical example 2 and numerical example 1 in the absence of public intervention, and when the subsidy (either *ad valorem* or specific) is given only to residents. The equivalent subsidy is computed in order to imply either the same price paid by residents that the *ad valorem* one—specific subsidy (1)—or, alternatively, the same public expenditure: specific subsidy (2).

Table 5.5 Effects of the *ad valorem* and specific subsidy only for residents for different non-resident's demand functions ($\alpha = 0.5$ and $\sigma = 0.75$)

	Low price-elastic non-residents' demand (numerical example 1)				High price-elastic non-residents' demand (numerical example 2)			
	Without subsidies	*Ad valorem*	Specific (1)	Specific (2)	Without subsidies	*Ad valorem*	Specific (1)	Specific (2)
p_S	73.06	107.08	86.28	88.91	34.17	38.33	36.40	36.53
p_d^R	73.06	26.77	26.77	17.56	34.17	9.58	9.58	8.21
ΔCS^R	–	11.59	11.59	15.17	–	9.61	9.61	10.23
ΔCS^{NR}	–	– 15.94	– 7.06	– 8.33	–	– 1.43	– 0.82	– 0.86
ΔPS	–	25.96	20.73	25.32	–	3.55	3.39	3.59
ΔGS	–	– 29.41	– 21.79	– 29.41	–	– 12.99	– 12.12	– 12.99

Note: σ denotes the *ad valorem* subsidy and α the percentage of resident passengers. p_S is the price charged by the airline and p_d^R the price paid by resident passengers. ΔCS^R is the change in resident consumers' surplus, ΔCS^{NR} the change in non-resident consumers' surplus, ΔPS the change in the airline's profits, and ΔGS the change in taxpayers' surplus
Specific (1): the specific subsidy that yields the same price for residents that the *ad valorem* one
Specific (2): the specific subsidy that yields the same government's expenditure that the *ad valorem* one
Prices are in euros, while changes in consumers' surplus, the airline's profits, and government's expenditure are measured in millions of euros

In the case in which non-residents' demand is very price-elastic (numerical example 2), the increase of the ticket-price charged by the airline is relatively small both with an *ad valorem* or the equivalent specific subsidy for residents. Still, with the specific subsidy (1) residents are indifferent, but non-residents and taxpayers are better-off than with the *ad valorem* subsidy. Similarly, with the specific subsidy (2) the public expenditure is the same, but residents and non-residents are better-off than with the *ad valorem* subsidy. Therefore, besides the proportion of resident passengers on the route, non-residents' demand price-elasticity is a variable that may mitigate the monopolist's market power and, thus, the differences between the *ad valorem* subsidy for residents and the equivalent specific one.

From these numerical simulations, we can conclude that the economic evaluation of any subsidy aimed for residents has to be performed taking into account the particular characteristics of the route. The economic consequences of choosing the wrong mechanism are greater the higher the proportion of resident passengers on the route is, and the less important is the non-residents' market for airlines. Moreover, the higher the level of competition on the route, the more effective the subsidy (either *ad valorem* or specific) aimed exclusively for residents is. Thus, competition should be promoted on the route as much as possible, either by encouraging the entry of more airlines in the route, by promoting intermodal competition (e.g., between air and maritime transport), or by removing all possible barriers to airlines' entry.

5.6 Conclusions

In this chapter, we have shown that the public policy objective of promoting transport connectivity and fair treatment of people living in disadvantaged areas, through subsidization of airfares, should be implemented with a specific subsidy. A fixed discount per trip and route is always superior to the *ad valorem* alternative except in the case of perfect competition.

When airlines have some market power, the *ad valorem* subsidy is equivalent to the combined use of a specific subsidy plus a profit subsidy. Hence, although a specific subsidy leads to the same price for residents as the *ad valorem* one, the *ad valorem* subsidy increases the monopolist's market power and, thus, implies that the airline charges a higher price, resulting in higher profits for the airline, and higher government's expenditure.

Moreover, with the *ad valorem* subsidy, the higher the ticket-price the higher the subsidy. When airlines have market power, this implies that: First, with the *ad valorem* subsidy they do not have enough incentives to be cost efficient and charge lower prices. Second, with the *ad valorem* subsidy, airlines do not have enough incentives to invest in cost reduction technologies. Third, with the *ad valorem* subsidy the more cost-inefficient is the airline, the higher the subsidy is. Under imperfect competition, the fact that more inefficient firms receive higher subsidies implies unfair competition. Fourth, the higher the proportion of resident passengers in the route, the more difficult is price competition. The reason is that, due to the *ad valorem* subsidy, while resident passengers only benefit from a proportion of any price reduction, airlines bear the whole cost of such a decrease (the proportion that otherwise would pay residents and the proportion that otherwise would pay the government). This fact discourages price competition among airlines. Finally, with the *ad valorem* subsidy passengers that buy more expensive tickets receive a higher subsidy.

The difference between the *ad valorem* subsidy for residents and the specific one is higher, the higher the proportion of resident passengers on the route is, and the less important is the non-residents' demand for airlines (either because their WTP is low or because they are very price-inelastic). Any CBA to evaluate the effects of the subsidy has to be performed route by route, taking into account the particular characteristics of the route and the period of time. Empirical models that use aggregate data are not informative enough to distinguish those routes where the policy is being effective from those routes where the policy is producing important non-desirable effects in the market.

Appendix 1: Mathematical Expressions of the Model

Equilibrium in the Absence of Public Subsidies: The Monopoly Case

The monopoly carrier chooses the ticket-price p_S that solves the following maximization program:

$$\underset{P_S}{\text{Max}}\, \alpha N(p_S - c)x^R + (1 - \alpha)N(p_S - c)x^{NR}, \tag{5.23}$$

with $x^R = \frac{a_R - P_S}{b_R}$, and $x^{NR} = \frac{a_{NR} - P_S}{b_{NR}}$.

The first-order condition of the above maximization program is given by:

$$\frac{N}{b_R b_{NR}}(b_{NR}\alpha(a_R + c) + b_R(1 - \alpha)(a_{NR} + c) - 2p_S(\alpha b_{NR} + (1 - \alpha)b_R)) = 0. \tag{5.24}$$

Solving the first-order condition we obtain the following optimal ticket-price:

$$p_S^0 = \frac{b_{NR}\alpha(a_R + c) + b_R(1 - \alpha)(a_{NR} + c)}{2(\alpha b_{NR} + (1 - \alpha)b_R)}. \tag{5.25}$$

By substituting the optimal ticket-price in the demand function of each passenger, we obtain the following demanded quantities per passenger (residents and non-residents, respectively):

$$x_0^R = \frac{a_R - p_S^0}{b_R} = \frac{\alpha a_R b_{NR} + 2a_R b_R(1 - \alpha) - (1 - \alpha)b_R a_{NR} - c(\alpha b_{NR} + (1 - \alpha)b_R)}{2b_R(\alpha b_{NR} + (1 - \alpha)b_R)},$$

$$x_0^{NR} = \frac{a_{NR} - p_S^0}{b_{NR}} = \frac{(1 - \alpha)b_R a_{NR} + 2\alpha a_{NR}b_{NR} - \alpha a_R b_{NR} - c(\alpha b_{NR} + (1 - \alpha)b_R)}{2b_{NR}(\alpha b_{NR} + (1 - \alpha)b_R)}. \tag{5.26}$$

Equilibrium with Ad Valorem Subsidies for Residents: The Monopoly Case

When the route is operated by a monopolist and an *ad valorem* subsidy only for residents is introduced, the airline solves the following maximization program:

$$\underset{P_S}{\text{Max}}\, \alpha N(p_S - c)x^R + (1 - \alpha)N(p_S - c)x^{NR}, \tag{5.27}$$

where x^R and x^{NR} represent, given the price that they finally pay, the quantity demanded by residents and non-residents, respectively. Thus, $x^R = \frac{a_R - (1 - \sigma)p_S}{b_R}$ and $x^{NR} = \frac{a_{NR} - p_S}{b_{NR}}$.

The first-order condition of the above maximization program is given by:

$$\frac{N}{b_R b_{NR}}(\alpha b_{NR}(-2p_S(1 - \sigma) + c(1 - \sigma) + a_R) + (1 - \alpha)b_R(-2p_S + c + a_{NR})) = 0. \tag{5.28}$$

Solving the first-order condition we obtain the following optimal ticket-price:

$$p_S^1 = \frac{\alpha b_{NR}(c(1 - \sigma) + a_R) + (1 - \alpha)b_R(c + a_{NR})}{2(b_R(1 - \alpha) + \alpha b_{NR}(1 - \sigma))}. \tag{5.29}$$

The ticket-price finally paid by residents with an *ad valorem* subsidy only for residents is given by:

$$p_d^R = p_S^1(1 - \sigma) = \frac{\alpha b_{NR}(c(1 - \sigma) + a_R) + (1 - \alpha)b_R(c + a_{NR})}{2(b_R(1 - \alpha) + \alpha b_{NR}(1 - \sigma))}(1 - \sigma). \quad (5.30)$$

By substituting the optimal ticket-price in the demand function of each passenger, we obtain the following demanded quantities per passenger (residents and non-residents, respectively):

$$x_1^R = \frac{a_R - p_S^1(1 - \sigma)}{b_R}$$
$$= \frac{\alpha b_{NR}(1 - \sigma)(-c(1 - \sigma) + a_R) + (1 - \alpha)b_R(-c(1 - \sigma) + 2a_R - a_{NR}(1 - \sigma))}{2b_R(b_R(1 - \alpha) + \alpha b_{NR}(1 - \sigma))},$$

$$x_1^{NR} = \frac{a_{NR} - p_S^1}{b_{NR}}$$
$$= \frac{\alpha b_{NR}(-c(1 - \sigma) + 2a_{NR}(1 - \sigma) - a_R) + (1 - \alpha)b_R(a_{NR} - c)}{2b_{NR}(b_R(1 - \alpha) + \alpha b_{NR}(1 - \sigma))}. \quad (5.31)$$

Equilibrium with Specific Subsidies for Residents: The Monopoly Case

When the route is operated by a monopolist and a specific subsidy only for residents is introduced, the airline solves the following maximization program:

$$\underset{p_S}{\text{Max}} \, \alpha N(p_S - c)x^R + (1 - \alpha)N(p_S - c)x^{NR}, \quad (5.32)$$

where x^R and x^{NR} represents, given the price that they finally pay, the quantity demanded by residents and non-residents, respectively. Thus, $x^R = \frac{a_R - p_S + s}{b_R}$ and $x^{NR} = \frac{a_{NR} - p_S}{b_{NR}}$.

The first-order condition of the above maximization program is given by:

$$\frac{N}{b_R b_{NR}}(\alpha b_{NR}(-2p_S + s + c + a_R) + (1 - \alpha)b_R(-2p_S + c + a_{NR})) = 0. \quad (5.33)$$

Solving the first-order condition we obtain the following optimal ticket-price:

$$p_S^1 = \frac{\alpha b_{NR}(s + c + a_R) + (1 - \alpha)b_R(c + a_{NR})}{2(b_R(1 - \alpha) + \alpha b_{NR})}. \quad (5.34)$$

The ticket-price finally paid by residents with a specific subsidy only for residents is given by:

$$p_d^R = p_S^1 - s = \frac{\alpha b_{NR}(s + c + a_R) + (1 - \alpha)b_R(c + a_{NR})}{2(b_R(1 - \alpha) + \alpha b_{NR})} - s. \quad (5.35)$$

By substituting the optimal ticket-price in the demand function of each passenger, we obtain the following demanded quantities per passenger (residents and non-residents, respectively):

$$x_1^R = \frac{a_R - p_S^1 + s}{b_R} = \frac{\alpha b_{NR}(s - c + a_R) + (1 - \alpha)b_R(-c + 2s + 2a_R - a_{NR})}{2b_R(b_R(1 - \alpha) + \alpha b_{NR})},$$

$$x_1^{NR} = \frac{a_{NR} - p_S^1}{b_{NR}} = \frac{\alpha b_{NR}(-s - c + 2a_{NR} - a_R) + (1 - \alpha)b_R(a_{NR} - c)}{2b_{NR}(b_R(1 - \alpha) + \alpha b_{NR})}. \quad (5.36)$$

Appendix 2: Some Empirical Evidence: The Case of Spain

Table 5.6 *Ad valorem* subsidy on flights from/to the Canary Islands from July 2018 to June 2019: interislands routes

Route (direct flights)	Total passenger-trips	Percentage of residents (%)	Percentage of the discount (%)	Average price paid by non-residents (€)	Average price paid by residents (€)
Gran Canaria–Tenerife	1,129,230	85.5	75	59.19	14.80
Gran Canaria–Lanzarote	804,732	88.9	75	69.03	17.26
Tenerife North–La Palma	760,709	88.9	75	55.61	13.90
Gran Canaria–Fuerteventura	657,273	89.2	75	61.89	15.47
Tenerife North–Lanzarote	382,369	88.2	75	92.72	23.18
Tenerife North–Fuerteventura	288,186	90.0	75	86.67	21.67
Tenerife North–El Hierro	205,221	82.1	75	62.60	15.65
Gran Canaria–La Palma	157,285	87.4	75	89.23	22.31
Tenerife North–La Gomera	60,693	77.4	75	46.21	11.55
Gran Canaria–El Hierro	49,352	85.8	75	86.79	21.70

(continued)

Table 5.6 (continued)

Route (direct flights)	Total passenger-trips	Percentage of residents (%)	Percentage of the discount (%)	Average price paid by non-residents (€)	Average price paid by residents (€)
Gran Canaria–La Gomera	8584	90.9	75	64.48	16.12
Total interisland routes	4,503,731	87.56	75	66.79	16.70

Source: de Rus & Socorro (2022)

Table 5.7 *Ad valorem* subsidy on flights from/to the Balearic Islands from July 2018 to June 2019: Interislands routes

Route (direct flights)	Total passenger-trips	Percentage of residents (%)	Percentage of the discount (%)	Average price paid by non-residents (€)	Average price paid by residents (€)
Mallorca–Ibiza	535,191	75.5	75	64.68	16.17
Mallorca–Menorca	376,274	80.2	75	61.99	15.50
Menorca–Ibiza	4364	28.7	75	84.93	21.23
Total interisland routes	915,829	77.21	75	63.56	16.39

Source: de Rus & Socorro (2022)

Table 5.8 *Ad valorem* subsidy on flights from/to the Canary Islands from July 2018 to June 2019: domestic non-interislands routes

Route (direct flights)	Total passenger-trips	Percentage of residents (%)	Percentage of the discount (%)	Average price paid by non-residents (€)	Average price paid by residents (€)
Gran Canaria–Madrid	1,656,580	45.4	75	119.73	29.93
Gran Canaria–Sevilla	219,716	54.6	75	113.33	28.33
Gran Canaria–Málaga	184,476	50.0	75	98.24	24.56
Gran Canaria–Granada	33,957	69.4	75	93.03	23.26
Gran Canaria–Alicante	26,484	41.9	75	155.59	38.90
Gran Canaria–Valencia	71,042	45.2	75	92.41	23.10
Gran Canaria–Barcelona	508,768	42.7	75	107.13	26.78
Gran Canaria–Bilbao	111,313	40.1	75	134.49	33.62
Gran Canaria–Santander	9049	31.0	75	75.04	18.76
Gran Canaria–Asturias	22,159	52.8	75	144.52	36.13

(continued)

Table 5.8 (continued)

Route (direct flights)	Total passenger-trips	Percentage of residents (%)	Percentage of the discount (%)	Average price paid by non-residents (€)	Average price paid by residents (€)
Gran Canaria–Vigo	20,047	78.6	75	195.96	48.99
Gran Canaria–Santiago	98,581	45.6	75	101.51	25.38
Gran Canaria–A Coruña	11,056	36.8	75	141.83	35.46
Tenerife–Madrid	1,827,992	39.7	75	124.39	31.10
Tenerife–Sevilla	287,136	46.8	75	118.44	29.61
Tenerife–Málaga	245,471	37.8	75	98.15	24.54
Tenerife–Granada	28,213	58.4	75	87.95	21.99
Tenerife–Alicante	99,597	30.9	75	96.40	24.10
Tenerife–Valencia	96,969	30.7	75	93.81	23.45
Tenerife–Barcelona	684,484	33.4	75	116.56	29.14
Tenerife–Zaragoza	19,709	34.7	75	130.71	32.68
Tenerife–Bilbao	199,522	27.2	75	126.12	31.53
Tenerife–Asturias	73,242	32.6	75	141.99	35.50
Tenerife–Vigo	20,682	55.1	75	181.93	45.48
Tenerife–Santiago	167,620	36.0	75	110.71	27.68
Tenerife–A Coruña	10,975	24.5	75	135.20	33.80
Lanzarote–Madrid	489,925	29.3	75	113.84	28.46
Lanzarote–Sevilla	79,668	38.3	75	74.11	18.53
Lanzarote–Málaga	33,839	34.1	75	96.77	24.19
Lanzarote–Valencia	47,866	23.6	75	72.33	18.08
Lanzarote–Barcelona	158,766	23.5	75	132.64	33.16
Lanzarote–Bilbao	107,396	14.1	75	129.91	32.48
Lanzarote–Asturias	25,076	22.9	75	144.39	36.10
Lanzarote–Santiago	84,168	30.3	75	84.87	21.22
Fuerteventura–Madrid	329,297	25.6	75	117.33	29.33
Fuerteventura–Sevilla	52,888	35.1	75	66.88	16.72
Fuerteventura–Málaga	30,929	33.7	75	83.33	20.83
Fuerteventura–Valencia	4998	27.2	75	100.73	25.18
Fuerteventura–Barcelona	161,431	19.2	75	98.91	24.73
Fuerteventura–Bilbao	32,515	13.7	75	132.52	33.13
Fuerteventura–Santiago	34,981	43.9	75	121.31	30.33
La Palma–Madrid	137,671	29.4	75	155.67	38.92
Total domestic non-interisland routes	8,546,254	38.3	75	116.83	29.21

Source: de Rus & Socorro (2022)

Table 5.9 *Ad valorem* subsidy on flights from/to the Balearic Islands from July 2018 to June 2019: domestic non-interislands routes

Route (direct flights)	Total passenger-trips	Percentage of residents (%)	Percentage of the discount (%)	Average price paid by non-residents (€)	Average price paid by residents (€)
Mallorca–Sevilla	323,796	48.2	75	90.36	22.59
Mallorca–Malaga	206,954	38.6	75	79.27	19.82
Mallorca–Granada	188,003	60.8	75	85.33	21.33
Mallorca–Alicante	278,252	40.1	75	77.29	19.32
Mallorca–Valencia	511,361	40.2	75	69.33	17.33
Mallorca–Barcelona	2,093,583	36.6	75	55.49	13.87
Mallorca–Zaragoza	82,346	36.6	75	81.69	20.42
Mallorca–Madrid	1,977,045	38.6	75	79.84	19.96
Mallorca–Bilbao	244,409	36.4	75	87.87	21.97
Mallorca–Santiago	153,813	44.0	75	90.13	22.53
Menorca–Madrid	290,011	21.9	75	99.67	24.92
Menorca–Valencia	51,731	18.6	75	89.71	22.43
Menorca–Barcelona	823,028	26.8	75	70.23	17.56
Ibiza–Sevilla	86,532	30.9	75	65.17	16.29
Ibiza–Málaga	132,643	24.7	75	60.93	15.23
Ibiza–Alicante	83,635	24.4	75	75.41	18.85
Ibiza–Valencia	324,009	27.4	75	47.15	11.79
Ibiza–Madrid	816,174	24.8	75	64.47	16.12
Ibiza–Barcelona	1,191,357	22.3	75	60.61	15.15
Ibiza–Bilbao	94,823	14.9	75	66.76	16.69
Total domestic non-interisland routes	9,953,505	33.4	75	71.76	17.94

Source: de Rus & Socorro (2022)

References

Adler, M. D., & Posner, E. A. (2001). Implementing cost–benefit analysis when preferences are distorted. In M. D. Adler & E. A. Posner (Eds.), *Cost–benefit analysis: Legal, economic and philosophical perspectives*. University of Chicago Press.

AIReF. (2020). *Spending review Fase II. Estudio Infraestructuras de Transporte*. Report available online at https://www.airef.es/wp-content/uploads/2020/07/INFRAESTRUCTURAS/ESTUDIO_INFRAESTRUCTURAS_SPENDINGREVIEW.pdf

Albalate, D., & Bel, G. (Eds.). (2017). *Evaluating high-speed rail: Interdisciplinary perspectives*. Routledge, Taylor & Francis Group.

Becker, G. S. (1965). A theory of the allocation of time. *The Economic Journal, 75*(299), 493–517.

Beria, P., de Rus, G., Quinet, E., & Schulz, C. (2012). A comparison of rail liberalisation levels across four European countries. *Research in Transportation Economics, 36*(1), 110–120.

Campos, J., & de Rus, G. (2009). Some stylized facts about high speed rail. A review of HSR experiences around the world. *Transport Policy, 16*, 19–28.

Campos, J., de Rus, G., & Barron, I. (2009). The cost of building and operating a new high speed rail line. In G. de Rus (Ed.), *Economic analysis of high speed rail in Europe*. BBVA Foundation. Available at: https://www.fbbva.es/wpcontent/uploads/2017/05/dat/DE_2012_economic_analysis_high_speed_rail.pdf.

Campos, J., & Gagnepain, P. (2009). Measuring the intermodal effects of high speed rail. In G. de Rus (Ed.), *Economic analysis of high speed rail in Europe*. BBVA Foundation. Available at: https://www.fbbva.es/wpcontent/uploads/2017/05/dat/DE_2012_economic_analysis_high_speed_rail.pdf.

Chetty, R. (2009). Sufficient statistics for welfare analysis: A bridge between structural and reduced-form methods. *Annual Review of Economics, 1*, 451–487.

Collier, P., & Venables, A. J. (2018). Who gets the urban surplus? *Journal of Economic Geography, 18*(3), 523–538.

de Rus, G. (2012). *Economic evaluation of the high speed rail*. Ministry of Finance of Sweden, Expert Group on Environmental Studies.

de Rus, G. (2021a). *Introduction to cost–benefit analysis: Looking for reasonable shortcuts* (2nd ed.). Edward Elgar.

de Rus, G. (2021b). The economic rationale for high-speed rail. In M. Börjesson & R. Vickerman (Eds.), *International encyclopedia of transportation* (Vol. 1, pp. 419–424). Elsevier.

de Rus, G. (2023). *Cost–benefit analysis for the social appraisal of projects*. European Investment Bank Institute.

de Rus, G., Campos, J., & Nombela, G. (2003). *Economía del Transporte*. Antoni Bosch.

© The Editor(s) (if applicable) and The Author(s), under exclusive license to Springer Nature Switzerland AG 2023
G. de Rus et al., *Economic Evaluation of Transport Projects*,
https://doi.org/10.1007/978-3-031-35959-0

de Rus, G., & Johansson, P. O. (2019). *Measuring the economic effects of transport improvements* (Policy Papers 2019/01). FEDEA, Madrid.

de Rus, G., & Nash, C. (2009). In what circumstances is investment in HSR worthwhile? In de Rus, G. (Ed.), *Economic analysis of high speed rail in Europe* (pp. 51–70). BBVA Foundation. Available at: https://www.fbbva.es/wpcontent/uploads/2017/05/dat/DE_2012_economic_analysis_high_speed_rail.pdf.

de Rus, G., & Socorro, M. P. (2017). Planning, evaluation and financing of transport infrastructures: Rethinking the basics. *Review of Network Economics, 16*(2), 143–160.

de Rus, G., & Socorro, M. P. (2019). Pricing and investment in alternative transport infrastructures. *Transportation Research Part A: Policy and Practice, 119*, 96–107.

de Rus, G., & Socorro, M. P. (2022). Subsidies in air transport markets: The economic consequences of choosing the wrong mechanism. *Transportation Research Part E: Logistic and Transportation Review, 160*, 1–15.

de Rus, G., Socorro, M. P., Valido, J., & Campos, J. (2022). Cost–benefit analysis of transport projects: Theoretical framework and practical rules. *Transport Policy, 123*, 25–39.

DeSerpa, A. C. (1971). A theory of the economics of time. *The Economic Journal, 81*, 828–846.

Dupuit, J. (1844). On the measurement of the utility of public works. *International Economic Papers, 2*, 83–110.

Duranton, G., & Puga, D. (2004). Micro-foundations of urban agglomeration economies. In J. V. Henderson & J.-F. Thisse (Eds.), *Handbook of urban and regional economics. Cities and Geography* (Vol. 4, pp. 2063–2117). Elsevier.

EIB. (2023). *The economic appraisal of investment projects at the EIB* (2nd edn). European Investment Bank. Available at www.eib.org/en/publications/

Engel, E., Fischer, R., & Galetovic, A. (2014). *The economics of public-private partnerships*. Cambridge University Press.

Gomez-Ibañez, J. A., & de Rus, G. (2006). *Competition in the railway industry: An international comparative analysis*. Edward Elgar.

Graham, D. J. (2007). Agglomeration, productivity and transport investment. *Journal of Transport Economics and Policy, 41*(3), 317–343.

Graham, D. J., & Gibbons, S. (2019). Quantifying wider economic impacts of agglomeration for transport appraisal: Existing evidence and future directions. *Economics of Transportation, 19*, 1–21.

Grob, L., & Craven, N. (2017). *Analysis of regional differences in global rail projects by cost, length and project stage* (UIC Working Paper). Available at https://uic.org

Harberger, A. C. (1965). *Survey of literature on cost–benefit analysis for industrial project evaluation*. Paper prepared for the Inter-Regional Symposium in Industrial Project Evaluation. Reprinted in Harberger, A. C. *Project Evaluation (Collected papers) (Midway reprint series)*. The University of Chicago Press.

Hensher, D. A. (2011). Valuation of travel time savings. In A. de Palma, R. Lindsey, E. Quinet, & R. Vickerman (Eds.), *A handbook of transport economics*. Edward Elgar Publishing.

Jara-Díaz, S., Astroza, S., Bhat, C. R., & Castro, M. (2016). Introducing relations between activities and goods consumption in microeconomic time use models. *Transportation Research Part B: Methodological, 93*, 162–180.

Johansson, P. O. (1993). *Cost–benefit analysis of environmental change*. Cambridge University Press.

Johansson, P. O., & de Rus, G. (2018). *Evaluating large projects when there are substitutes: Looking for possible shortcuts* (Working paper, 2018/12). FEDEA, Madrid.

Johansson, P. O., & de Rus, G. (2019). On the treatment of foreigners and foreign-owned firms in the cost–benefit analysis of transport projects. *Journal of Transport Economics and Policy, 53*(3), 275–287.

Johansson, P. O., & Kriström, B. (2016). *Cost–benefit analysis for project appraisal*. Cambridge University Press.

Just, R. E., Hueth, D. L., & Schmitz, A. (2004). *The welfare economics of public policy: A practical approach to project and policy evaluation*. Edward Elgar.

Kleven, H. J. (2018). *Sufficient statistics revisited* (NBER working paper).

Kriström, B. (2023). *East is east and west is west? A gentle introduction to links between CGE and CBA*. European Investment Bank Institute.

Krugman, P. (1998). Space: The Final Frontier. *Journal of Economic Perspectives, 12*(2), 161–174.

Laird, J. J., Nash, C., & Mackie, P. (2014). Transformational transport infrastructure: Cost–benefit analysis challenges. *The Town Planning Review, 85*(6), 709–730.

Laird, J. J., & Venables, A. J. (2017). Transport investment and economic performance: A framework for project appraisal. *Transport Policy, 56*, 1–11.

Mackie, P., Graham, D. J., & Laird, J. (2011). The direct and wider impacts of transport projects: A review. In A. de Palma, R. Lindsey, E. Quinet, & R. Vickerman (Eds.), *A handbook of transport economics*. Edward Elgar.

Mackie, P., Jara-Díaz, S., & Fowkes, A. S. (2001). The value of travel time savings in evaluation. *Transportation Research Part E, 37*, 91–106.

Mackie, P., & Preston, J. (1998). Twenty-one sources of errors and bias in transport appraisal. *Transport Policy, 5*, 1–7.

Mackie, P., Worsley, T., & Eliasson, J. (2014). Transport appraisal revisited. *Research in Transportation Economics, 47*, 3–18.

Mas-Colell, A., Whinston, M. D., & Green, J. R. (1995). *Microeconomic theory*. Oxford University Press.

Mohring, H. (1971). Alternative welfare gains and loss measures. *Western Economic Journal, 9*, 349–368.

Mohring, H. (1993). Maximizing, measuring, and *not* double counting transportation improvement-benefits: A primer on closed-and open-economy cost–benefit analysis. *Transportation Research B, 27B*(6), 413–424.

Mohring, H., Jr., & Williamson, H. F. (1969). Scale and 'industrial reorganisation' economies of transport improvements. *Journal of Transport Economics and Policy*, 251–271.

Nash, C. A. (2014). Appraisal in the rail sector: General issues and British experience in dealing with them. *Research in Transportation Economics, 47*, 19–26.

Neuberger, H. (1971). User benefit in the evaluation of transport and land use plans. *Journal of Transport Economics and Policy, 3*, 52–75.

Oum, T. H., & Waters, W. G. I. I. (1997). Recent developments in cost function research in transportation. In G. De Rus & C. Nash (Eds.), *Recent developments in transport economics* (pp. 33–73). Ashgate Publishing.

Preston, J. (2013). *The economics of investment in high speed rail* (Discussion paper no. 2013-30). International Transport Forum. OECD.

Preston, J. (2017). Direct and indirect effects of high-speed rail. In D. Albalate, & G. Bel (Eds.), *Evaluating high-speed rail: Interdisciplinary perspectives*. Routledge, Taylor & Francis Group.

Rodríguez-Pose, A. (2018). The revenge of the places that don't matter (and what to do about it). *Cambridge Journal of Regions, Economy and Society, 11*, 189–209.

Small, K. A. (1999). Project evaluation. In J. A. Gómez-Ibáñez, W. B. Tye, & W. Clifford (Eds.), *Essays in transportation economics*. Brooking Institution Press.

UIC. (2018). *High speed rail: fast track to sustainable mobility*. International Railways Union. Available at www.uic.org

UIC. (2022). *Atlas high-speed rail 2022*. International Union of Railways. Available at https://uic.org/passenger/highspeed/article/high-speed-data-and-atlas

Venables, A. J. (2007). Evaluating urban transport improvements: Cost–benefit analysis in the presence of agglomeration and income taxation. *Journal of Transport Economics and Policy, 41*, 173–188.

Venables, A. (2019). Transport appraisal: Wider economic benefits. In *CBA workshop on the assessment of large-cross border transport projects*. Innovation and Networks Executive Agency (INEA). European Commission.

Willig, R. D. (1976). Consumer's surplus without apology. *American Economic Review, 66*(4), 589–597.

Zerbe, R. O. (2018). The concept of standing in benefit–cost analysis. In R. O. Zerbe (Ed.), *Teaching benefit–cost analysis* (pp. 58–68). Edward Elgar.

The manufacturer's authorised representative in the EU is Springer
Nature Customer Service Centre GmbH, Europaplatz 3, 69115 Heidelberg,
Germany. If you have any concerns regarding our products, please
contact ProductSafety@springernature.com

Printed and bound by CPI Group (UK) Ltd, Croydon, CR0 4YY

29/04/2026

02099530-0002